Friedrich Hirzebruch
Thomas Berger
Rainer Jung

Manifolds and Modular Forms

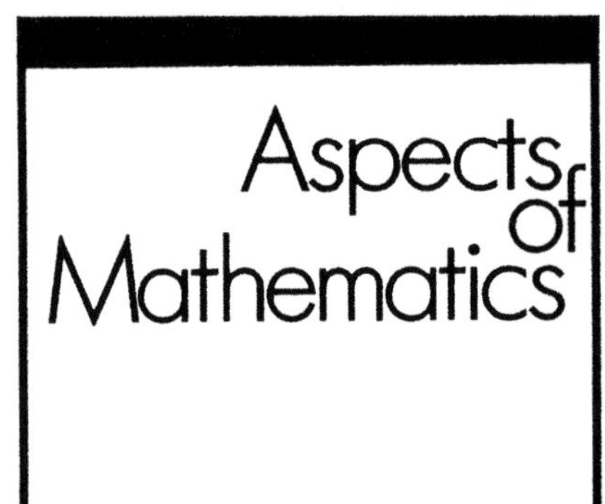

Edited by Klas Diederich

*A Publication of the Max-Planck-Institut für Mathematik, Bonn

Volumes of the German-language subseries "Aspekte der Mathematik" are listed at the end of the book.

Friedrich Hirzebruch
Thomas Berger
Rainer Jung

Manifolds and Modular Forms

Translated by Peter S. Landweber

A Publication from the
Max-Planck-Institut für Mathematik, Bonn

Second Edition

Professor Dr. Friedrich Hirzebruch,
Thomas Berger, and Rainer Jung
Max-Planck-Institut für Mathematik
Gottfried-Claren-Str. 26
53225 Bonn
Germany

First Edition 1992
Second Edition 1994

Appendix III: *Elliptic genera of level N for complex manifolds* reprinted with permission of Kluwer Academic Publishers

Mathematics Subject Classification: 57-02, 11F11, 33C45, 33E05, 55N22, 55R10, 57R20, 58G10

Cover design: Wolfgang Nieger, Wiesbaden
Typeset using ArborText Publisher and $\mathrm{T_{E}X}$

ISSN 0179-2156
ISBN 978-3-528-16414-0   ISBN 978-3-663-10726-2 (eBook)
DOI 10.1007/978-3-663-10726-2

Preface

During the winter term 1987/88 I gave a course at the University of Bonn under the title "Manifolds and Modular Forms". I wanted to develop the theory of "Elliptic Genera" and to learn it myself on this occasion. This theory due to Ochanine, Landweber, Stong and others was relatively new at the time. The word "genus" is meant in the sense of my book "Neue Topologische Methoden in der Algebraischen Geometrie" published in 1956: A genus is a homomorphism of the Thom cobordism ring of oriented compact manifolds into the complex numbers. Fundamental examples are the signature and the $\hat{A}$-genus. The $\hat{A}$-genus equals the arithmetic genus of an algebraic manifold, provided the first Chern class of the manifold vanishes. According to Atiyah and Singer it is the index of the Dirac operator on a compact Riemannian manifold with spin structure.

The elliptic genera depend on a parameter. For special values of the parameter one obtains the signature and the $\hat{A}$-genus. Indeed, the universal elliptic genus can be regarded as a modular form with respect to the subgroup $\Gamma_0(2)$ of the modular group; the two cusps giving the signature and the $\hat{A}$-genus.

Witten and other physicists have given motivations for the elliptic genus by theoretical physics using the free loop space of a manifold. This led Witten to conjectures concerning the rigidity of the elliptic genus generalizing a result of Atiyah and myself on the rigidity of the equivariant $\hat{A}$-genus:

If the circle S^1 acts on a spin manifold, then the equivariant $\hat{A}$-genus is a formal difference of two finite dimensional vector spaces (positive and negative harmonic spinors) which are representation spaces of S^1. Our theorem (published in 1970) states that the difference of the characters of these representations (a finite Laurent series) is a constant, indeed equal to zero if the action is non-trivial. The proof uses the Atiyah-Singer equivariant index and fixed point theorem applied to the S^1-action.

The rigidity theorem in the case of the equivariant universal elliptic genus of spin manifolds with S^1-action was first proved by Taubes and then by Bott and Taubes. I did not know it during the time of my course. I stressed special cases (strict multiplicativity in certain fibre bundles) following the pioneers Ochanine and Landweber.

Shortly after my course the proceedings of a Princeton meeting (1986) edited by Landweber became available. They give an excellent description of the history of the young theory and the state of the art around the time of my course.

During the second part of the course (changing the original program) I began to develop the theory of elliptic genera of level N for complex manifolds, finding out that the differential equations for the characteristic functions are given by polynomials which I called almost-Chebyshev. These are polynomials such that the critical values except one are all equal up to sign. A little later I learned from Th. J. Rivlin that such polynomials were already well-known under the name Zolotarev polynomials. The universal elliptic genus of level N is a modular form for $\Gamma_1(N)$, the values in the cusps being certain holomorphic Euler numbers (if the first Chern class of the manifold is divisible by N). In this way the universal elliptic genus of level N is very close to the Riemann-Roch theorem formulated and proved in my book mentioned above.

During the winter term 1987/88 I was occasionally away on official business. My course was taken over several times by N.-P. Skoruppa who gave a thorough and expert presentation of results on modular forms needed for the theory. P. Baum was visiting the Max-Planck-Institut. He also lectured in my course and gave an introduction to the Dirac operator. In this way the first two of the four appendices of this book came into existence.

The course was directed to students after their Vordiplom. Some of them had basic knowledge on characteristic classes. However, I tried to explain without proofs all the prerequisite material to make the course understandable to a large audience: characteristic classes, cobordism theory, Atiyah-Singer index theorem, Riemann-Roch theorem, etc.

Thomas Berger and Rainer Jung belonged to the students and prepared notes. This was a considerable job, because I did not have a usable manuscript. The result of their efforts were inofficial lecture notes in German which became ready in July 1988. P. Landweber used these notes for a course at Rutgers University in the spring of 1989 and prepared a translation into English for his students; in addition he proposed numerous corrections and improvements. Thus the present English text materialized. For the German version of the notes also Gottfried Barthel and Michael Puschnigg contributed improvements.

The book has two further appendices:

In August 1987 the sixteenth of the well-known and highly estimated International Conferences on Differential Geometrical Methods in Theoretical Physics organized by Konrad Bleuler took place in Como. I was invited to write up my talk for the proceedings. The talk was a general survey on elliptic genera. Instead of writing this up, I wrote a paper "Elliptic genera of level N for complex manifolds" which is reproduced here as Appendix III, incorporating some corrections and improvements by Thomas Berger.

A rigidity theorem for the level N genus is contained in the paper. It was proved during my visit in Cambridge (England) in March 1988 when I was a guest of Robinson College. When Michael Atiyah came to Bonn in February 1988 he explained to me Bott's approach to the rigidity for spin manifolds mentioned above and indicated its relationship to our old paper on the $\hat{A}$-genus. (The Atiyah-Singer equivariant index and fixed point theorem is used.) We had further discussions in Oxford in March 1988, before I proved the result for the level N case (consulting Bott's Cargèse lectures). Later the paper by Bott and Taubes became available.

The role of the Zolotarev polynomials was already pointed out. At the end of my paper I announced the plan to write a separate paper on these polynomials and their role for the elliptic genus. But, indeed, Rainer Jung took over. This became the topic of his Diplomarbeit, in which he carried out my plan and proved several other interesting theorems. His Appendix IV is a condensed version of the Diplomarbeit.

I wish to thank all the mathematicians mentioned above for their cooperation and help. I want to express special thanks to my coauthors Thomas Berger and Rainer Jung who wrote the original notes and to whom the book owes its existence. Many thanks are due to Peter Landweber for the translation and for all mathematical help. I am grateful that Nils-Peter Skoruppa and Paul Baum added their lectures to the book, and I must thank again Rainer Jung for adding his Diplomarbeit and for much work during the final stages of the preparation of the manuscript. Many thanks are due also to Mrs. Iris Abdel Hafez of the Max-Planck-Institut for her excellent work in typesetting the manuscript.

Bonn   F. Hirzebruch
March 23, 1992

Preface to the second edition

The text remained almost unchanged. Only some slight obscurities and several misprints have been corrected.

The authors would like to express thanks to all readers who have contributed their comments or will do so in the future. We are also indebted to the computing centre of the Max-Planck-Institut managed by Sven Maurmann for support during the preparation of both editions of this book.

Bonn   F. Hirzebruch
September 1993   Th. Berger
                                                                                              R. Jung

Notice to the reader

The special sign "⧉" after a proof indicates the end of that proof. If it occurs at the end of a stated theorem, proposition, lemma or corollary no proof for that statement will be given.

Contents

1 Background

1.1 Cobordism Theory

R. Thom [Th54] and L.S. Pontrjagin introduced the concept of cobordism, in which equivalence classes of manifolds are provided with a ring structure.

Let V be a compact, oriented, differentiable, n-dimensional manifold without boundary. For an oriented manifold W with boundary, ∂W denotes its boundary with the induced orientation.

Definition: *A manifold V **bounds** (V **is bounded**) in case there exists a compact, oriented, differentiable, $(n+1)$-dimensional manifold W so that $\partial W = V$.*

Example: S^n bounds ($S^n = \partial B^{n+1}$).

Further manifolds which are boundaries are the complex projective spaces of odd dimension. In order to show this, we use the following

Lemma: *There exists a fibration $\sigma : P_{2k+1}(\mathbb{C}) \to P_k(\mathbb{H})$ with fibre $P_1(\mathbb{C})$. (Here $\mathbb{H}$ denotes the skew-field of quaternions.)*

Proof: The bijection $\widehat{\sigma} : \mathbb{C}^{2k+2} \to \mathbb{H}^{k+1}$,

$$\widehat{\sigma}(z_1, \ldots, z_{2k+2}) = (z_1 + z_2 \cdot j, \ldots, z_{2k+1} + z_{2k+2} \cdot j),$$

is compatible with left multiplication by complex scalars, and thereby induces the desired fibration σ.
The vectors $(z_1, \ldots, z_{2k+2})$ and $(\bar{z}_2, -\bar{z}_1, \ldots, \bar{z}_{2k+2}, -\bar{z}_{2k+1})$ span a plane in $\mathbb{C}^{2k+2}$, i.e. their images generate a projective line in $P_{2k+1}(\mathbb{C})$. Since $-j \cdot (z_1 + z_2 \cdot j) = \bar{z}_2 - \bar{z}_1 \cdot j$ holds in $\mathbb{H}$, this line is mapped to a point of $P_k(\mathbb{H})$ under σ, and on dimensional grounds is precisely the fibre. $\qquad\qquad\Box$

Proposition: $P_{2k+1}(\mathbb{C})$ *bounds.*

Idea of proof: The above fibration yields a locally trivial fibre bundle with fibre $P_1(\mathbb{C}) \cong S^2$, which one can give the structure group $\mathrm{SO}(3)$. One can extend its operation to the associated disk bundle (with fibre the 3-ball). This is then a manifold of dimension $4k + 3$, whose boundary is exactly $P_{2k+1}(\mathbb{C})$. $\qquad\qquad\Box$

Definition: *Two manifolds V and W are called **cobordant** ($V \sim W$) in case the manifold $V + (-W)$ bounds. Here "$+$" denotes the disjoint union, and $-W$ is the manifold W with reversed orientation.*

Remark: The relation $\sim$ is an equivalence relation. The reflexivity follows from $V + (-V) = \partial(V \times I)$, where I denotes the unit interval. It is elementary to verify the symmetry; for transitivity, one must glue two manifolds along diffeomorphic boundary components.

Definition: *Let Ω^n denote the set of equivalence classes of n-dimensional, compact, oriented, differentiable manifolds with respect to $\sim$.*

Remark: The pair $(\Omega^n, +)$ is a finitely generated abelian group. The Cartesian product of manifolds induces a map $\Omega^n \times \Omega^m \to \Omega^{n+m}$. Thereby $\Omega = \sum_{n=0}^{\infty} \Omega^n$ becomes a graded commutative ring with 1, the so-called cobordism ring. Thus, one has $\alpha \cdot \beta = (-1)^{n \cdot m} \beta \cdot \alpha$ for $\alpha \in \Omega^n$, $\beta \in \Omega^m$; the equivalence class of a point is the unit 1.

Theorem (Thom): *$\Omega^n \otimes \mathbb{Q} = 0$ for $4 \nmid n$ and Ω^{4k} is a finitely generated abelian group with rank equal to the number of partitions of k.* 📖

Hence $\Omega \otimes \mathbb{Q}$ is a commutative ring.

Definition: *A sequence M^1, $M^2, \ldots$ of manifolds (with $\dim M^i = 4i$) is called a* **basis sequence** *(of the cobordism ring) if $\Omega \otimes \mathbb{Q} = \mathbb{Q}[M^1, M^2, \ldots]$, where the M^i are considered as equivalence classes.*

Theorem: *The spaces $P_2(\mathbb{C})$, $P_4(\mathbb{C})$, $P_6(\mathbb{C}), \ldots$ form a basis sequence of $\Omega \otimes \mathbb{Q}$, i.e.*

$$\Omega \otimes \mathbb{Q} = \mathbb{Q}[P_2(\mathbb{C}), P_4(\mathbb{C}), P_6(\mathbb{C}), \ldots]$$

as a graded polynomial ring, where $P_{2k}(\mathbb{C})$ has weight (or dimension) $4k$. 📖

Example:

$$\Omega^2 \otimes \mathbb{Q} = \langle P_2(\mathbb{C}) \rangle,$$
$$\Omega^4 \otimes \mathbb{Q} = \langle P_2(\mathbb{C}) \times P_2(\mathbb{C}), P_4(\mathbb{C}) \rangle,$$
$$\Omega^6 \otimes \mathbb{Q} = \langle P_2(\mathbb{C}) \times P_2(\mathbb{C}) \times P_2(\mathbb{C}), P_4(\mathbb{C}) \times P_2(\mathbb{C}), P_6(\mathbb{C}) \rangle.$$

1.2 Characteristic classes

Theorem: *For each complex vector bundle E over a manifold X there exist Chern classes $c_i(E)$ with:*

1) $c_i(E) \in H^{2i}(X; \mathbb{Z})$, $c_0(E) = 1$,

   $c(E) := \sum_{i=0}^{\infty} c_i(E) \in H^*(X; \mathbb{Z})$,

2) $c_i(f^* E) = f^* c_i(E)$,

3) $c(E \oplus F) = c(E) \cdot c(F)$,

4) $c(H) = 1 - g$.

The Chern classes are uniquely determined by the properties 1) *to* 4).

Remarks:

On 1): One has $c_i(E) = 0$ for $i > \mathrm{rk}\, E$, and $c(E)$ is called the total Chern class of E.

On 2): Let $f : Y \to X$ be a continuous map. Then f^*E denotes the pulled-back bundle. This has as total space

$$\widetilde{E} = \{(y,e) \in Y \times E \mid \pi(e) = f(y)\},$$

and one has the following diagram

$$\begin{array}{ccc} f^*E & \overset{pr_2}{\to} & E \\ {\scriptstyle pr_1} \downarrow & & \downarrow {\scriptstyle \pi} \\ Y & \overset{f}{\to} & X \end{array}$$

where pr_i is the projection onto the i-th component. On the right side of the equation f^* denotes the usual mapping $f^* : H^*(X;\mathbb{Z}) \to H^*(Y;\mathbb{Z})$.

On 3): Let E and F be two vector bundles over X. Then $E \oplus F$ is the Whitney sum of the bundles. The fibre $(E \oplus F)_x$ over a point $x \in X$ is the direct sum of the fibres E_x and F_x of E, resp. F. The multiplication rule therefore gives:

$$c_1(E \oplus F) = c_1(E) + c_1(F),$$
$$c_2(E \oplus F) = c_2(E) + c_1(E) \cdot c_1(F) + c_2(F) \quad \text{etc.}$$

On 4): Here $g \in H^2(P_n(\mathbb{C});\mathbb{Z})$ denotes the generating element of the cohomology ring of $P_n(\mathbb{C})$, Poincaré dual to the homology class of the hyperplane $P_{n-1}(\mathbb{C}) \subset P_n(\mathbb{C})$. The bundle H is the Hopf bundle or tautological bundle over $P_n(\mathbb{C})$, having as fibre over each point of $P_n(\mathbb{C})$ the line in $\mathbb{C}^{n+1}$ represented by it.

A differentiable manifold is called almost complex if its tangent bundle is derived from a complex bundle (by forgetting the complex structure).

Theorem: *Let X be an almost complex manifold of complex dimension n. Then the n-th Chern class $c_n = c_n(TX) \in H^{2n}(X;\mathbb{Z})$ satisfies $c_n[X] = e(X)$, where $e(X)$ is the Euler number of X (and $c_n[X]$ denotes the evaluation of c_n on the fundamental cycle of the oriented manifold X).*

By the total Chern class of an almost complex manifold one means the total Chern class of its tangent bundle. With this convention we formulate the

Lemma: $c(P_n(\mathbb{C})) = (1 + g)^{n+1} = 1 + (n + 1) \cdot g + \binom{n+1}{2} \cdot g^2 + \cdots + (n + 1) \cdot g^n$.

Proof: Exercise, apply the above theorem (cf. [MiSt74], §14).

Definition: *Let E be a real vector bundle over X. Then we define the **Pontrjagin** classes $p_i(E)$ by*

$$p_i(E) := (-1)^i c_{2i}(E \otimes \mathbb{C}) \in H^{4i}(X;\mathbb{Z}).$$

Proposition: *Let E be a real vector bundle over X of rank n. Then the Pontrjagin classes satisfy the following axioms:*

1) $p_i(E) \in H^{4i}(X; \mathbb{Z})$, $p_0(E) = 1$,

   $p(E) := \sum_{i=0}^{\infty} p_i(E) \in H^*(X; \mathbb{Z})$,

2) $p_i(f^*E) = f^* p_i(E)$,
3) $p(E \oplus F) = p(E) \cdot p(F)$ *modulo 2-torsion*,
4) $p(H_{\mathbb{R}}) = 1 + g^2$.

Remarks:

On 1): One has $p_i(E) = 0$ for $2i > n$.
On 3): "modulo 2-torsion" means $2 \cdot (p(E \oplus F) - p(E) \cdot p(F)) = 0$.
On 4): For a complex bundle F, we denote by $F_{\mathbb{R}}$ the underlying real bundle, i.e. F as a real bundle when one forgets the complex vector space structure on the fibres.

Proof of the proposition: 1) and 2) are clear.
Since the complex bundle $E \otimes \mathbb{C}$ comes from a real bundle, one has $E \otimes \mathbb{C} \cong \overline{E \otimes \mathbb{C}}$. As we shall see in section 1.5, this implies for the Chern classes that

$$c_{2i+1}(E \otimes \mathbb{C}) = c_{2i+1}\left(\overline{E \otimes \mathbb{C}}\right) = (-1)^{2i+1} c_{2i+1}(E \otimes \mathbb{C})$$
$$= -c_{2i+1}(E \otimes \mathbb{C})$$
$$\Rightarrow \quad 2c_{2i+1}(E \otimes \mathbb{C}) = 0.$$

From this 3) is immediate.
Property 4) holds, since for a complex vector bundle E always $E_{\mathbb{R}} \otimes \mathbb{C} \cong E \oplus \overline{E}$. This is clear, since one can always conjugate a matrix of the form $\left(\begin{smallmatrix} \mathrm{Re}\,(z) & -\mathrm{Im}\,(z) \\ \mathrm{Im}\,(z) & \mathrm{Re}\,(z) \end{smallmatrix}\right)$ into the form $\left(\begin{smallmatrix} z & 0 \\ 0 & \bar{z} \end{smallmatrix}\right)$ canonically. Hence for a complex vector bundle E one has

$$\sum_{i=0}^{\infty} (-1)^i p_i(E) = c(E) \cdot \sum_{i=0}^{\infty} (-1)^i c_i(E).$$

For the Hopf bundle H this means:

$$1 - p_1(H_{\mathbb{R}}) = (1 + c_1(H)) \cdot (1 - c_1(H))$$
$$= (1 - g) \cdot (1 + g) = 1 - g^2$$
$$\Rightarrow \quad p(H_{\mathbb{R}}) = 1 + p_1(H_{\mathbb{R}}) = 1 + g^2.$$

Analogously, one obtains for the projective spaces, noting that $(1 + g)^{n+1} \cdot (1 - g)^{n+1} = (1 - g^2)^{n+1}$, the result:

$$p(P_n(\mathbb{C})) = (1 + g^2)^{n+1}.$$

Definition: *Let X be a compact, oriented, almost complex manifold of dimension $2n$ and $(i_1, \ldots, i_r)$ a partition of n, i.e. $\sum_{j=1}^{r} i_j = n$. Then the **Chern number** corresponding to this partition is defined as*

$$\left(\prod_{j=1}^{r} c_{i_j}(X) \right) [X].$$

*In the differentiable case, one defines **Pontrjagin numbers** for a manifold of dimension $4n$ analogously as*

$$\left(\prod_{j=1}^{r} p_{i_j}(X) \right) [X].$$

As we shall see in section 1.6, the Pontrjagin numbers are cobordism invariants. More precisely, there holds the following

Theorem (Thom): *Two manifolds represent the same class in $\Omega \otimes \mathbb{Q}$ precisely when all their Pontrjagin numbers coincide.*   □

1.3 Pontrjagin classes of quaternionic projective spaces

Analogous to the cell decomposition of complex projective space $P_n(\mathbb{C}) = \mathbb{C}^n \cup P_{n-1}(\mathbb{C})$, one obtains a cell decomposition of $P_n(\mathbb{H})$ into cells of dimension $0, 4, 8, \ldots, 4n$. There further holds $H^*(P_n(\mathbb{H}); \mathbb{Z}) = \mathbb{Z}[u]/(u^{n+1})$, where u is a generator of $H^4(P_n(\mathbb{H}); \mathbb{Z})$.

Theorem (Hirzebruch):

$$\begin{aligned}
p(P_n(\mathbb{H})) &= (1+u)^{2n+2} \cdot (1+4u)^{-1} \\
&= (1+u)^{2n+2} \cdot \left(1 - 4u + 16u^2 - 64u^3 \pm \cdots \right).
\end{aligned}$$

Proof (cf. [Hi53]): Let $\sigma : P_{2k+1}(\mathbb{C}) \to P_k(\mathbb{H})$ be the fibre bundle of section 1.1 with fibre $P_1(\mathbb{C})$. Now (by the Leray-Hirsch theorem) $\sigma^* : H^{4r}(P_k(\mathbb{H}); \mathbb{Z}) \to H^{4r}(P_{2k+1}(\mathbb{C}); \mathbb{Z})$ is an isomorphism for all r. We choose u so that $\sigma^*(u) = g^2 \in H^4(P_{2k+1}(\mathbb{C}); \mathbb{Z})$; then $u^k[P_k(\mathbb{H})] = 1$ determines the orientation of $P_k(\mathbb{H})$. The tangent bundle $\widetilde{T}$ of the projective space $P_{2k+1}(\mathbb{C})$ splits into the direct sum of the sub-bundle E along the fibres of the map σ and the pullback of the tangent bundle T of the projective space $P_k(\mathbb{H})$:

$$\widetilde{T} \cong E \oplus \sigma^* T$$

$$\Rightarrow \quad \left(1 + g^2 \right)^{2k+2} = p(\widetilde{T}) = p(E) \cdot \sigma^* p(T).$$

The bundle E is orientable, and so comes from a complex line bundle ($SO(2) \cong U(1)$). Notice that the restriction of this complex bundle to a fibre is $TP_1(\mathbb{C})$. The map $i : P_1(\mathbb{C}) \to P_{2k+1}(\mathbb{C})$ induces on the cohomology level the identity in dimensions $0, 1$ and 2, otherwise is the zero map. To see this, one considers a cell decomposition of $P_{2k+1}(\mathbb{C})$ into affine subspaces, which contains a cell decomposition of the fibre $P_1(\mathbb{C})$. Hence

$$i^* c_1(E) = c_1(i^* E) = c_1(P_1(\mathbb{C})).$$

Since $c_1(P_1(\mathbb{C}))[P_1(\mathbb{C})] = e(S^2) = 2$, we have

$$
\begin{aligned}
c_1(E) &= 2g \\
\Rightarrow \quad c(E) &= 1 + 2g \\
\Rightarrow \quad 1 - p_1(E) &= (1 + 2g) \cdot (1 - 2g) \\
\Rightarrow \quad p(E) &= 1 + p_1(E) = 1 + 4g^2.
\end{aligned}
$$

Therefore

$$
\begin{aligned}
\left(1 + g^2\right)^{2k+2} &= \left(1 + 4g^2\right) \cdot \sigma^*(p(T)) \\
\Rightarrow \quad \sigma^*(p(T)) &= \left(1 + g^2\right)^{2k+2} \cdot \left(1 + 4g^2\right)^{-1}.
\end{aligned}
$$

Since σ^* is an isomorphism in all dimensions $4r$, we have

$$p(T) = (1 + u)^{2k+2} \cdot (1 + 4u)^{-1}.$$

Example: The first Pontrjagin class of $P_n(\mathbb{H})$ is $p_1(P_n(\mathbb{H})) = (2n + 2) \cdot u - 4 \cdot u = (2n - 2) \cdot u$.

For $n = 1$ it follows that $p_1 = 0$, hence $p(P_1(\mathbb{H})) = p(S^4) = 1$.

Definition: *A bundle E is called **stably trivial** if a trivial bundle F exists so that $E \oplus F$ is also trivial.*

Remark: For a stably trivial bundle E one has $p(E) = 1$. For

$$
\begin{aligned}
1 = c((E \oplus F) \otimes \mathbb{C}) &= c(E \otimes \mathbb{C}) \cdot c(F \otimes \mathbb{C}) = c(E \otimes \mathbb{C}) \\
\Rightarrow \quad p_i(E) &= (-1)^i c_{2i}(E \otimes \mathbb{C}) = 0 \qquad \text{for } i = 1, \dots, n.
\end{aligned}
$$

Example: The normal bundle NS^k of the sphere S^k is trivial, and also $TS^k \oplus NS^k = T\mathbb{R}^{k+1}|_{S^k}$. Therefore the tangent bundle of S^k is stably trivial. Hence we have also given a direct calculation of the Pontrjagin classes of $P_1(\mathbb{H}) = S^4$.

For even k, there exists no nowhere vanishing vector field on S^k (for k even, the Euler number is $e(S^k) = 2 \neq 0$). Therefore the tangent bundle is then non-trivial. For odd k, this is also valid for $n \neq 1, 3, 7$, although difficult (cf. [Mi58] and [AtHi61b]). These three exceptional cases correspond to the existence of the real division algebras $\mathbb{C}$, $\mathbb{H}$ and the Cayley numbers.

Remark: The Pontrjagin classes are fixed under each automorphism of a differentiable manifold, since an automorphism of the tangent bundle is induced.

Example: For $k > 1$ there is no automorphism of $P_k(\mathbb{H})$ carrying u to $-u$, since $p_1(P_k(\mathbb{H})) = (2k - 2) \cdot u \neq 0$. For $k = 1$ there exists the antipodal map $S^4 \to S^4$, $x \mapsto -x$.

1.4 Characteristic classes and invariants

The following is intended to provide motivation for why the Chern resp. Pontrjagin classes will be regarded formally as elementary symmetric functions, and why they yield important information on the isomorphism class of a bundle. For a more detailed treatment of this so-called splitting principle we refer to section 4.4.

A complex line bundle over X has structure group $\mathbb{C}^*$; the first Chern class $c_1 \in H^2(X;\mathbb{Z})$ yields an isomorphism from the group of isomorphism classes of line bundles with tensor product as group operation onto the second integral cohomology: Let E and F be two line bundles over X, and $X = \bigcup_i U_i$ a common system of local trivializations with change of coordinates e_{ij}, resp. $f_{ij} : U_i \cap U_j \to \mathbb{C}^*$. Then (p, z) in the chart $U_j \times \mathbb{C}$ is to be identified with $(p, e_{ij} \cdot z)$, resp. $(p, f_{ij} \cdot z)$ in $U_i \times \mathbb{C}$, and the e_{ij}, resp. f_{ij} satisfy the cocycle condition $e_{ij} \cdot e_{jk} = e_{i\kappa}$, resp. $f_{ij} \cdot f_{j\kappa} = f_{i\kappa}$. Because of the commutativity of the structure group $\mathbb{C}^*$, the products $e_{ij} \cdot f_{ij}$ also satisfy the cocycle condition, and so define a line bundle $E \otimes F$, the tensor product of E and F.

Let now $E = E_1 \oplus \cdots \oplus E_n$ be a direct sum of complex line bundles, and let $x_i = c_1(E_i) \in H^2(X;\mathbb{Z})$. The complex vector bundle $E_1 \oplus \cdots \oplus E_n$ has rank n and total Chern class

$$c(E) = 1 + c_1 + \cdots + c_n = (1 + x_1) \cdots (1 + x_n).$$

Therefore $c_r = \sigma_r(x_1, \ldots, x_n)$ is the r-th elementary symmetric function in the x_i (by a comparison of dimensions of cohomology classes).

If one introduces a Hermitian metric on E_i, then the structure group of E_i can be reduced to $\mathrm{U}(1) \cong S^1 = T^1$; that of their direct sum E can therefore be reduced to

$$T^n = S^1 \times \cdots \times S^1 = \left\{ A \in \mathrm{U}(n) \mid A = \begin{pmatrix} e^{2\pi i \varphi_1} & & \\ & \ddots & \\ & & e^{2\pi i \varphi_n} \end{pmatrix}, \varphi_i \in \mathbb{R} \right\}.$$

This torus $T^n \subset \mathrm{U}(n)$ is maximal in the compact connected Lie group $\mathrm{U}(n)$. Let G be a compact connected Lie group, then there holds for the maximal tori in G the following

Theorem: *Each element of G lies in a maximal torus. If T, T' are maximal tori, then there exists a $g \in G$ so that $gTg^{-1} = T'$ (cf. [HoSa40]).*   ⌑

This theorem is a generalization of the principal axis theorem of linear algebra: since each element lies in some maximal torus, it can be conjugated into the standard maximal torus T^n.

An element $g \in \mathrm{U}(n)$ with $gTg^{-1} = T$ transports the different S^1-components of the torus into one another, and so yields an isomorphism of the $\mathrm{U}(n)$-bundle which permutes the summands E_i, but need not be an isomorphism of T^n-bundles. Hence, each expression invariant under $\mathrm{U}(n)$-isomorphism must remain fixed under permutations of the x_i, i.e. is a polynomial in the Chern classes c_j, since these are the elementary symmetric functions $\sigma_j(x_1, \ldots, x_n)$ in the x_i.

For Pontrjagin classes analogously holds: Let E be a real vector bundle of rank $2n$ and with structure group $\mathrm{SO}(2n)$. The standard maximal torus is

$$T^n = \left\{ A \in \mathrm{SO}(2n) \mid A = \begin{pmatrix} R(2\pi\varphi_1) & & \\ & \ddots & \\ & & R(2\pi\varphi_n) \end{pmatrix}, \varphi_i \in \mathbb{R} \right\},$$

where $R(2\pi\varphi) = \begin{pmatrix} \cos(2\pi\varphi) & -\sin(2\pi\varphi) \\ \sin(2\pi\varphi) & \cos(2\pi\varphi) \end{pmatrix}$. Assume that the structure group is reduced to $T^n \subset \mathrm{SO}(2n)$. Then we know that the vector bundle E is a direct sum of two-dimensional bundles E_i with structure group $\mathrm{SO}(2) \cong \mathrm{U}(1) \cong S^1$, thus $E = E_1 \oplus \cdots \oplus E_n$, and we define $x_i := c_1(E_i)$.

Elements $g \in \mathrm{SO}(2n)$ with $gTg^{-1} = T$ yield arbitrary exchanges of the E_i with an even number of orientation changes of the E_i, without altering the isomorphism type of the $\mathrm{SO}(2n)$-bundle E. Thus, the ring of invariant polynomials has generators $\sigma_j(x_1^2, \ldots, x_n^2)$ and $x_1 \cdots x_n$. Since

$$p(E) = p(E_1) \cdots p(E_n) = (1 + x_1^2) \cdots (1 + x_n^2),$$

$p_j = \sigma_j(x_1^2, \ldots, x_n^2)$ and $e = x_1 \cdots x_n$ generate the ring of invariants (e is called the Euler class and one has $e(TX)[X] = e(X)$, the Euler number of the manifold X).

Now let the rank of E be $2n+1$, the structure group $\mathrm{SO}(2n + 1)$, with maximal torus

$$T^n = \left\{ A \in \mathrm{SO}(2n + 1) \mid A = \begin{pmatrix} R(2\pi\varphi_1) & & & \\ & \ddots & & \\ & & R(2\pi\varphi_n) & \\ & & & 1 \end{pmatrix}, \varphi_i \in \mathbb{R} \right\}.$$

In this case an odd number of orientation changes of the E_i is also possible, because an odd permutation can be compensated by -1 in the lower corner of the permutation matrix. In odd dimensions the Euler class is therefore no longer invariant; the Pontrjagin classes $p_j = \sigma_j(x_1^2, \ldots, x_n^2)$ generate the ring of invariant polynomials.

Each polynomial which is symmetric in the x_i^2 can be expressed in the p_j; for example,

$$\sum_{i=1}^{n} x_i^2 = p_1, \qquad \sum_{i=1}^{n} x_i^4 = p_1^2 - 2p_2.$$

1.5 Representations and vector bundles

Let E be a complex vector bundle of rank n over a differentiable manifold X. Assume that the structure group of E is reduced to $U(n)$. Let ρ be a unitary representation of $U(n)$ of dimension m, so that $\rho : U(n) \to U(m)$ is a homomorphism. The group $U(n)$ then acts on $\mathbb{C}^m$ by means of ρ. We consider the vector bundle ρE, associated to E, with fibre $\mathbb{C}^m$.

Examples: By suitable choices of ρ one obtains for ρE the bundles:

1) $E^* \cong \overline{E}$ with rank n for $\rho(U) = \overline{U}$,
2) $\Lambda^k E$ with rank $\binom{n}{k}$ for $\rho(U)(v_1 \wedge \cdots \wedge v_k) = Uv_1 \wedge \cdots \wedge Uv_k$ where $v_1 \wedge \cdots \wedge v_k \in \Lambda^k \mathbb{C}^n$,
3) $S^k E$ with rank $\binom{n+k-1}{k}$ for $\rho(U)(v_1 \otimes \cdots \otimes v_k) = Uv_1 \otimes \cdots \otimes Uv_k$ where $v_1 \otimes \cdots \otimes v_k \in S^k \mathbb{C}^n$.

Let the bundle E have Chern classes $c_j(E) \in H^{2j}(X;\mathbb{Z})$. We want to express the Chern classes of the bundle ρE in terms of the $c_j(E)$. For this, we consider a formal factorization of the total Chern class

$$c(E) = 1 + c_1(E) + \cdots + c_n(E) = (1 + x_1) \cdots (1 + x_n),$$

i.e. $c_j(E)$ is the j-th elementary symmetric function $\sigma_j(x_1,\ldots,x_n)$ in the x_i. All expressions which are symmetric in the formal roots x_i of $c(E)$ are well-defined, since they can be expressed in terms of the $c_j(E)$. Thus, we can define the Chern character $\mathrm{ch}\,(E)$ of the bundle E as

$$\mathrm{ch}\,(E) := \sum_{i=1}^{n} e^{x_i}$$

$$= n + \sum_{i=1}^{n} x_i + \frac{1}{2}\sum_{i=1}^{n} x_i^2 + \frac{1}{6}\sum_{i=1}^{n} x_i^3 + \ldots$$

$$= n + c_1 + \frac{c_1^2 - 2c_2}{2} + \frac{c_1^3 - 3c_1c_2 + 3c_3}{6} + \cdots \in H^*(X;\mathbb{Q}).$$

Conversely, $\mathrm{ch}\,(E)$ determines the rational Chern classes of E (c_j can be expressed in terms of the cohomology components of dimension up to $2j$ of $\mathrm{ch}\,(E)$). For a trivial bundle E of rank n, one has $\mathrm{ch}\,(E) = n$. The Chern character of E has two beautiful properties:

$$\mathrm{ch}\,(E_1 \oplus E_2) = \mathrm{ch}\,(E_1) + \mathrm{ch}\,(E_2) \qquad \text{and} \qquad \mathrm{ch}\,(E_1 \otimes E_2) = \mathrm{ch}\,(E_1) \cdot \mathrm{ch}\,(E_2).$$

For a representation $\rho : U(n) \to U(m)$, we define its character $\mathrm{char}\,(\rho) : U(n) \to \mathbb{C}$ as the trace of ρA: $\mathrm{char}\,(\rho)(A) := \mathrm{tr}\,(\rho A)$ for all $A \in U(n)$. Now the standard maximal tori in $U(n)$, resp. $U(m)$, are

$$T = \left\{ A \in U(n) \mid A = \begin{pmatrix} e^{2\pi i x_1} & & \\ & \ddots & \\ & & e^{2\pi i x_n} \end{pmatrix}, \, x_i \in \mathbb{R} \right\}$$

and

$$\tilde{T} = \{A \in \mathrm{U}(m) \mid A = \begin{pmatrix} e^{2\pi i y_1} & & \\ & \ddots & \\ & & e^{2\pi i y_m} \end{pmatrix}, \ y_i \in \mathbb{R}\}.$$

Since $\rho(T)$ is again abelian, and therefore lies in a maximal torus, and since any two maximal tori are conjugate, we can change ρ (by conjugation in $\mathrm{U}(m)$) to an equivalent representation for which $\rho(T) \subset \tilde{T}$. Since conjugation of transition functions carries a bundle to an isomorphic bundle, the isomorphism class of $\rho(E)$ does not change, nor do its Chern classes. In $\rho(T) \subset \tilde{T}$ there holds

$$y_r(x_1, \ldots, x_n) = \sum_{s=1}^{n} a_{rs} x_s \qquad \text{with } a_{rs} \in \mathbb{Z}.$$

One calls the $y_r = (a_{r1}, \ldots, a_{rn})$ the weights of the representation (interpreted as linear forms), and obtains for the character of the representation ρ:

$$\mathrm{char}\,(\rho)|_T = \sum_{r=1}^{m} e^{2\pi i y_r}.$$

For the bundle ρE associated to E then holds (cf. [BoHi58]) the

Theorem: $\mathrm{ch}\,(\rho E) = \sum_{r=1}^{m} e^{y_r} \qquad \text{and} \qquad c(\rho E) = \prod_{r=1}^{m} (1 + y_r).$

Remark: More precisely, in place of y_r one must write $y_r(x_1, \ldots, x_n)$, where again $c_j(E) = \sigma_j(x_1, \ldots, x_n)$.

Example: In case $E = L_1 \oplus \cdots \oplus L_n$ is a sum of line bundles, $x_i = c_1(L_i)$ and $y_r = \sum_{s=1}^{n} a_{rs} x_s$, one has:

$$\rho(E) = (L_1^{a_{11}} \otimes \cdots \otimes L_n^{a_{1n}}) \oplus \cdots \oplus (L_1^{a_{m1}} \otimes \cdots \otimes L_n^{a_{mn}})$$

$$\Rightarrow \quad c(\rho E) = \prod_{r=1}^{m} (1 + a_{r1} c_1(L_1) + \cdots + a_{rn} c_1(L_n))$$

$$= \prod_{r=1}^{m} (1 + y_r)$$

$$\Rightarrow \quad \mathrm{ch}\,(\rho E) = \sum_{r=1}^{m} e^{y_r}.$$

Remark: This is even a proof of the above theorem, if one uses the following meta-principle (the splitting principle, cf. section 4.4): All identities for Chern classes of vector bundles that one can prove by means of the formal factorization of the total Chern class are true.

Examples: For our special representations at the start of this section we obtain:

1) For $\rho E = E^*$:

$$y_i = -x_i \quad \text{and} \quad c(E^*) = \prod_{i=1}^{n}(1 - x_i) = \sum_{i=1}^{n}(-1)^i c_i(E).$$

2) For $\rho E = \Lambda^k E$:

$$y_{i_1,\ldots,i_k} = x_{i_1} + \cdots + x_{i_k} \quad \text{and} \quad c(\Lambda^k E) = \prod_{1 \leq i_1 < \cdots < i_k \leq n}(1 + (x_{i_1} + \cdots + x_{i_k})).$$

3) For $\rho E = S^k E$:

$$y_{i_1,\ldots,i_k} = x_{i_1} + \cdots + x_{i_k} \quad \text{and} \quad c(S^k E) = \prod_{1 \leq i_1 \leq \cdots \leq i_k \leq n}(1 + (x_{i_1} + \cdots + x_{i_k})).$$

4) For $\rho E = \Lambda^k E^*$:

$$y_{i_1,\ldots,i_k} = -(x_{i_1} + \cdots + x_{i_k}) \quad \text{and}$$
$$c(\Lambda^k E^*) = \prod_{1 \leq i_1 < \cdots < i_k \leq n}(1 - (x_{i_1} + \cdots + x_{i_k})).$$

The Chern character therefore satisfies the following formula:

$$\sum_{k=0}^{n} \mathrm{ch}(\Lambda^k E^*) \cdot t^k = \prod_{i=1}^{n}(1 + te^{-x_i}).$$

In view of the additivity and multiplicativity of the Chern character, we will also denote this expression by

$$\mathrm{ch}\Big(\sum_{k=0}^{\infty}(\Lambda^k E^*) \cdot t^k\Big).$$

With the abbreviation

$$\Lambda_t E := \sum_{k=0}^{\infty}(\Lambda^k E) \cdot t^k$$

we have more simply

$$\mathrm{ch}\,(\Lambda_t E^*) = \prod_{i=1}^{n}(1 + te^{-x_i}).$$

For a trivial bundle of rank n this is the well-known formula:

$$\sum_{k=0}^{n} \binom{n}{k} \cdot t^k = (1 + t)^n = \prod_{i=1}^{n} (1 + t).$$

Analogously, there holds for the symmetric powers, with the notation

$$S_t E := \sum_{k=0}^{\infty} (S^k E) \cdot t^k$$

the formula

$$\mathrm{ch}\,(S_t E^*) = \prod_{i=1}^{n} \frac{1}{1 - te^{-x_i}}\,.$$

For a trivial bundle this becomes

$$\sum_{k=0}^{\infty} \binom{n+k-1}{k} \cdot t^k = \sum_{k=0}^{\infty} \binom{-n}{k} \cdot (-1)^k t^k = (1 - t)^{-n} = \prod_{i=1}^{n} \frac{1}{1 - t}\,.$$

Now let E be a real vector bundle of rank $2n$ over X, having structure group reduced to $\mathrm{SO}(2n)$. The bundle E has characteristic classes $p_i(E) \in H^{4i}(X;\mathbb{Z})$ and $e(E) \in H^{2n}(X;\mathbb{Z})$.

To each homomorphism $\rho : \mathrm{SO}(2n) \to \mathrm{U}(m)$ one again obtains a bundle ρE associated to E, as well as by analogy to the above the character $\mathrm{char}\,(\rho)$ and the weights y_r with respect to the standard tori, which we have already considered earlier. In this situation we have the analog of the above theorem:

Theorem: $\mathrm{ch}\,(\rho E) = \sum_{r=1}^{m} e^{y_r}$ *and* $c(\rho E) = \prod_{r=1}^{m} (1 + y_r)$.

The total Pontrjagin class of $\Lambda^k E^*$ is calculated from the total Chern class of $\Lambda^k E^* \otimes \mathbb{C}$. This bundle is however canonically isomorphic to $\Lambda^k(E^* \otimes \mathbb{C})$. Since we already know the Chern classes of exterior powers, we need only consider the representation $\rho : \mathrm{SO}(2n) \subset \mathrm{U}(2n)$ with $\rho E = E \otimes \mathbb{C}$. For this representation, a box in the standard maximal torus of the form $R(2\pi x_r) = \begin{pmatrix} \cos(2\pi x_r) & -\sin(2\pi x_r) \\ \sin(2\pi x_r) & \cos(2\pi x_r) \end{pmatrix}$ goes over to the box $\begin{pmatrix} e^{2\pi i x_r} & 0 \\ 0 & e^{-2\pi i x_r} \end{pmatrix}$. The weights are therefore $\pm x_r$, and one obtains the

Theorem:

$$\mathrm{ch}\,(\Lambda_t E^* \otimes \mathbb{C}) = \mathrm{ch}\Big(\sum_{k=0}^{n} (\Lambda^k(E^* \otimes \mathbb{C})) \cdot t^k\Big)$$

$$= \prod_{i=1}^{n} \big((1 + te^{x_i}) \cdot (1 + te^{-x_i})\big).$$

Analogously,

$$\mathrm{ch}\left(S_t E^* \otimes \mathbb{C}\right) = \mathrm{ch}\left(\sum_{k=0}^{\infty} \left(S^k(E^* \otimes \mathbb{C})\right) \cdot t^k\right)$$
$$= \prod_{i=1}^{n} \frac{1}{(1 - te^{x_i})(1 - te^{-x_i})}.$$

1.6 Multiplicative sequences and genera

Definition: *Let R be an integral domain over $\mathbb{Q}$. Then a **genus** is a ring homomorphism $\varphi : \Omega \otimes \mathbb{Q} \to R$ with $\varphi(1) = 1$.*

Let $Q(x) = 1 + a_2 x^2 + a_4 x^4 + \cdots$ be an even power series, starting with 1 and with coefficients in R. For indeterminates x_i of weight two, where $1 \leq i \leq n$, the product $Q(x_1) \cdots Q(x_n)$ is symmetric in the x_i^2:

$$Q(x_1) \cdots Q(x_n) = 1 + a_2 \sum_{i=1}^{n} x_i^2 + \cdots .$$

The term of weight $4r$ can therefore be expressed as a homogeneous polynomial $K_r(p_1, \ldots, p_r)$ of weight $4r$ in the elementary symmetric functions p_j of the x_i^2, i.e.

$$Q(x_1) \cdots Q(x_n) = 1 + K_1(p_1) + K_2(p_1, p_2) + \cdots$$
$$+ K_n(p_1, \ldots, p_n) + K_{n+1}(p_1, \ldots, p_n, 0) + \cdots$$

where the polynomials K_r for $1 \leq r \leq n$ do not depend on n. The sequence $\{K_r\}$ of polynomials K_r is called the multiplicative sequence of polynomials associated to the power series $Q(x)$ (cf. [Hi56]). Now a genus $\varphi : \Omega \otimes \mathbb{Q} \to R$ is determined by the power series $Q(x)$, as follows:

Definition: *The **genus** φ_Q **corresponding to a power series** Q is defined, for every compact, oriented, differentiable manifold M of dimension $4n$, by*

$$\varphi_Q(M) := K_n(p_1, \ldots, p_n)[M] \in R$$

with $p_i = p_i(M) \in H^{4i}(M; \mathbb{Z})$. In addition, we put $\varphi_Q(M) := 0$ if $4 \nmid n$. With

$$K(TM) := K(p_1, \ldots, p_n) := 1 + K_1(p_1) + K_2(p_1, p_2) + \cdots$$

and $K(M) := K(TM)[M]$, we have $\varphi_Q(M) = K(M)$.

The genus belonging to a power series is therefore a linear combination of Pontrjagin numbers. In case it is clear from which power series Q the genus has been formed, we shall frequently write φ in place of φ_Q.

Lemma: *Such a φ_Q is a well-defined genus.*

Proof: To verify that φ is well-defined, it must be shown that φ vanishes on boundaries and is compatible with disjoint union and Cartesian product. Let $M^{4n} = \partial W^{4n+1}$. Since the normal bundle of M^{4n} in W^{4n+1} is trivial, the tangent bundle of W^{4n+1} restricted to the boundary is equal to the tangent bundle of M^{4n} plus a trivial line bundle. The Pontrjagin classes of M^{4n} are therefore those of W^{4n+1} restricted to the boundary (stability of Pontrjagin classes). Hence the Pontrjagin numbers of M^{4n} are equal to 0 (by Stokes's theorem).

The additivity of φ is clear. Now the multiplicative sequence $\{K_i\}$ possesses a multiplicative property (hence the name):

If p_i, p_i', p_i'' are indeterminates for which

$$1 + p_1 + p_2 + \cdots = (1 + p_1' + p_2' + \cdots) \cdot (1 + p_1'' + p_2'' + \cdots)$$

then one also has

$$\sum_{n \geq 0} K_n(p_1, \ldots, p_n) = \sum_{n \geq 0} K_n(p_1', \ldots, p_n') \cdot \sum_{n \geq 0} K_n(p_1'', \ldots, p_n'').$$

Exactly this assumption is fulfilled by the Pontrjagin classes of the cross product of two manifolds. As one now easily sees, φ is therefore also compatible with multiplication in the cobordism ring.

For an even normalized power series Q as above, let us now define $f(x) := x/Q(x)$. This is an odd power series, beginning with x and with coefficients in R (in the formation of $1/Q(x)$, one again obtains a formal power series with coefficients in R, since $Q(x)$ begins with 1).

Put $y = f(x)$ and $g = f^{-1}$ as the (formal) inverse function, i.e. $g(y) = x$. This power series g is also called the logarithm of the genus φ_Q (cf. section 3.1).

Lemma: $g'(y) = \displaystyle\sum_{n=0}^{\infty} \varphi_Q(P_n(\mathbb{C})) \cdot y^n.$

Proof:

$$c(P_n(\mathbb{C})) = (1 + x)^{n+1}, \quad x \in H^2(P_n(\mathbb{C}); \mathbb{Z}) \qquad (\text{earlier: } x = g)$$

$$\Rightarrow \quad p(P_n(\mathbb{C})) = \left(1 + x^2\right)^{n+1} = (1 + p_1)^{n+1}$$

$$\Rightarrow \quad K(p_1, \ldots, p_n) = K(p_1)^{n+1} = Q(x)^{n+1}.$$

We now calculate $\varphi_Q(P_n(\mathbb{C}))$ with the residue theorem and multiplicativity as follows:

$$\varphi_Q(P_n(\mathbb{C})) = \left(\frac{x}{f(x)}\right)^{n+1}[P_n(\mathbb{C})]$$

$$= \text{coefficient of } x^n \text{ in } \left(\frac{x}{f(x)}\right)^{n+1} = \text{res}_0\left(\frac{1}{f(x)}\right)^{n+1} dx$$

$$= \frac{1}{2\pi i}\int_\kappa \left(\frac{1}{f(x)}\right)^{n+1} dx = \frac{1}{2\pi i}\int_{f(\kappa)} \frac{1}{y^{n+1}}\, g'(y)\, dy$$

$$= \text{res}_0\left(\frac{g'(y)\, dy}{y^{n+1}}\right) = \text{coefficient of } y^n \text{ in } g'(y).$$

If f converges, κ corresponds to a circle about 0 in function theory; $f(\kappa)$ is then also a closed path with winding number one, since the power series f starts with x. If f doesn't converge the integral makes no sense but the substitution formula for residues of differentials holds for any formal power series f.   ⌸

On the other hand, if one prescribes the values of a genus φ on the complex projective spaces, therefore fixing the power series g', then one immediately obtains the series g and $f = g^{-1}$, and so also $Q(x) = x/f(x)$. Hence there is a one-to-one correspondence between genera and even power series $Q(x) = 1 + \cdots$. The power series Q is also called the characteristic power series of φ.

Example: Let $Q(x) = x/\tanh(x)$, so $f(x) = \tanh(x)$. Then $f'(x) = 1 - f(x)^2$, so $g'(y) = 1/(1 - y^2) = 1 + y^2 + y^4 + \cdots$. One thereby obtains a genus, taking the value 1 on all $P_{2k}(\mathbb{C})$, which is called the L-genus (cf. [Hi56]).

Convention: *Since we shall not distinguish between a genus and the corresponding multiplicative sequence, let us agree as follows: For a manifold M, the value of the genus will be denoted by $\varphi(M)$; for a bundle E over M, $\varphi(E)$ denotes the corresponding expression in the Pontrjagin classes of E. One therefore has $\varphi(M) = \varphi(TM)[M]$.*

1.7 Calculation of $\varphi(P_k(\mathbb{H}))$

Using the fibre bundle $\sigma : P_{2k+1}(\mathbb{C}) \to P_k(\mathbb{H})$ with fibre $P_1(\mathbb{C})$ and the identification of $g^2 \in H^4(P_{2k+1}(\mathbb{C}); \mathbb{Z})$ with $u \in H^4(P_k(\mathbb{H}); Z)$ by means of σ^*, we saw earlier that

$$p(P_k(\mathbb{H})) = \left(1 + g^2\right)^{2k+2} \cdot \left(1 + 4g^2\right)^{-1}.$$

Now let φ be the genus corresponding to some power series $Q(x) = x/f(x)$. Then $\varphi(P_k(\mathbb{H}))$ is the coefficient a_{2k} of x^{2k} in $\left(\frac{x}{f(x)}\right)^{2k+2} \cdot \left(\frac{f(2x)}{2x}\right)$. This is

$$a_{2k} = \text{res}_0 \frac{1}{f(x)^{2k+2}} \frac{f(2x)}{2}\, dx = \text{res}_0 \frac{1}{y^{2k+2}} \frac{f(2x)}{2}\, g'(y)\, dy.$$

There exists an even power series $\widetilde{h}(x)$ so that $f(2x) = 2f(x)f'(x)\widetilde{h}(x)$. Then $h(y) := \widetilde{h}(g(y))$ is an even power series with $f(2x) = 2f(x)f'(x)h(y)$, hence:

$$
\begin{aligned}
a_{2k} &= \mathrm{res}_0\, \frac{2f(x)}{2y^{2k+2}}\, f'(x)g'(y)h(y)\, dy \\
&= \mathrm{res}_0\, \frac{1}{y^{2k+1}}\, h(y)\, dy \qquad (\text{since } f(x) = y,\ f'(x) = 1/g'(y)) \\
&= \text{coefficient of } y^{2k} \text{ in } h(y).
\end{aligned}
$$

We have therefore proved the

Lemma: $\displaystyle\sum_{k=0}^{\infty} \varphi(P_k(\mathbb{H}))y^{2k} = h(y),\ \text{where}\ h(f(x)) = \frac{f(2x)}{2f(x)f'(x)}.$

Examples:

1) $Q(x) = \dfrac{x}{\sin(x)}$:

Then $f(x) = x/Q(x) = \sin(x)$ and $f(2x) = \sin(2x) = 2\sin(x)\cos(x) = 2f(x)f'(x)$, i.e. in this case $h \equiv 1$. The power series $f(x) = \sin(\alpha x)/\alpha$ are all those for which $h \equiv 1$ holds in the duplication formula. Therefore the corresponding genera φ vanish on all projective spaces $P_k(\mathbb{H})$ with $k > 0$. The genus corresponding to

$$
Q(x) = \frac{x/2}{\sinh(x/2)}
$$

is called the $\hat{A}$-genus, and the normalization is chosen so that for spin manifolds it is related to the index of an elliptic differential operator (the Dirac operator, cf. Appendix II).

2) $Q(x) = \dfrac{x}{\tanh(x)}$ (L-genus):

We have earlier calculated that $L(P_{2k}(\mathbb{C})) = 1$ for all k.

$$
\begin{aligned}
\tanh(2x) &= \frac{2\tanh(x)}{1 + \tanh(x)^2} = \frac{2\tanh(x)\big(1 - \tanh(x)^2\big)}{1 - \tanh(x)^4} \\
&= \frac{2\tanh(x)\,\tanh'(x)}{1 - \tanh(x)^4} \\
\Rightarrow \quad h(y) &= \frac{1}{1 - y^4} = 1 + y^4 + y^8 + \cdots \\
\Rightarrow \quad L(P_k(\mathbb{H})) &= \begin{cases} 0, & k \equiv 1\ (2), \\ 1, & k \equiv 0\ (2). \end{cases}
\end{aligned}
$$

One arrives at the so-called elliptic genera if one requires that in the duplication formula

$$h(y) = \frac{1}{1 - \varepsilon \cdot y^4} \, ,$$

which gives

$$\varphi(P_k(\mathbb{H})) = \begin{cases} 0, & k \equiv 1 \ (2), \\ \varepsilon^{k/2}, & k \equiv 0 \ (2). \end{cases}$$

The solution $f(x)$ to this kind of duplication formula is still not unique, rather it depends on a parameter δ. The latter is the value of the genus on $P_2(\mathbb{C})$, i.e. $g'(y) = 1 + \delta y^2 + \cdots$. These genera are called elliptic genera because of the connection with elliptic functions; more about this later.

Examples: $\delta = \varepsilon = 1$: L-genus; $\delta = -\frac{1}{8}, \varepsilon = 0$: $\hat{A}$-genus.

Since in dimension four there is only the Pontrjagin number p_1 and $p_1(P_2(\mathbb{C})) = 3$, in this dimension one has $L = p_1/3$ and $\hat{A} = -p_1/24$.

Exercise: Let $f(x) = x + \cdots$ be an odd power series, $g'(y)$ the derivative of the inverse function with $g'(y) = 1 + \delta \cdot y^2 + \cdots$, and suppose further that

$$f(2x) = \frac{2f(x)f'(x)}{1 - \varepsilon \cdot f(x)^4} \, .$$

For fixed δ, ε this power series is uniquely determined. Use this to prove that

$$f(u + v) = \frac{f(u)f'(v) + f'(u)f(v)}{1 - \varepsilon \cdot f(u)^2 f(v)^2} \, .$$

Deduce then that

$$f'^2 = 1 - 2\delta \cdot f^2 + \varepsilon \cdot f^4.$$

Now we can make the following

Definition: *A genus φ is called an **elliptic genus**, if the corresponding power series $f(x) = x/Q(x)$ satisfies one of the three equivalent conditions:*

1) $f'^2 = 1 - 2\delta \cdot f^2 + \varepsilon \cdot f^4,$

2) $f(u + v) = \dfrac{f(u)f'(v) + f'(u)f(v)}{1 - \varepsilon \cdot f(u)^2 f(v)^2},$

3) $f(2u) = \dfrac{2f(u)f'(u)}{1 - \varepsilon \cdot f(u)^4} \, .$

Examples: The two most important special cases are:

a) $\delta = \varepsilon = 1$: $f(x) = \tanh(x)$, $g'(y) = \frac{1}{1-y^2}$.

   1) $f'^2 = \left(1 - y^2\right)^2 = \left(1 - f^2\right)^2$.
   2)   Addition theorem for the hyperbolic tangent.

b) $\delta = -\frac{1}{8}$, $\varepsilon = 0$: $f(x) = 2\sinh(x/2)$, $g'(y) = \left(1 + \frac{1}{4}y^2\right)^{-1/2}$.

   1) $f'^2 = 1 + \frac{1}{4}y^2 = 1 + \frac{1}{4}f^2$.
   2)   Addition theorem for the hyperbolic sine.

1.8 Complex genera

Similar to the construction of the oriented cobordism ring, one can define relations between stably almost complex manifolds and so obtain the complex cobordism ring (cf. [No62]). For this ring, complex genera are again defined simply as ring homomorphisms. Just as in the oriented case, there is a one-to-one correspondence between genera and power series $Q(x) = 1 + b_1 x + \cdots$, but now these power series Q need no longer be even. Let X be a compact, almost complex manifold of real dimension $2n$. As usual, let

$$c(X) = 1 + c_1(X) + \cdots + c_n(X) = (1 + x_1) \cdots (1 + x_n).$$

Then the genus φ of X, belonging to the power series Q, is given by

$$\varphi(X) = \left(\prod_{i=1}^{n} Q(x_i)\right)[X].$$

Just as in the oriented case, one can associate to the power series Q a sequence of polynomials $K_r(c_1, \ldots, c_r)$. Here the polynomial K_r is homogeneous of degree r, if one assigns the degree j to the elementary symmetric function c_j in the x_i. These polynomials have the property that

$$\varphi(X) = K_n(c_1, \ldots, c_n)[X].$$

For reasons of degree, c_n enters only linearly into K_n. The coefficient of c_n is a fixed polynomial s_n in the coefficients b_j of the power series Q. In order to determine this polynomial, we shall make use of a trick. Since s_n only depends on the terms of degree at most n in the power series Q, we can write formally

$$1 + b_1 x + \cdots + b_n x^n \ = \ \prod_{i=1}^{n} (1 + \beta_i x)$$

$$\Rightarrow \quad \prod_{j=1}^{n} Q(x_j) \ \equiv \ \prod_{j=1}^{n}\prod_{i=1}^{n} (1 + \beta_i x_j) \qquad (x^{n+1})$$

$$\equiv \ \prod_{i=1}^{n}\Big(\prod_{j=1}^{n} (1 + \beta_i x_j)\Big) \qquad (x^{n+1})$$

$$\equiv \ \prod_{i=1}^{n} (1 + \beta_i c_1 + \cdots + \beta_i^n c_n) \qquad (x^{n+1}).$$

In order to determine s_n, we put $c_1 = \ldots = c_{n-1} = 0$. Since we must now consider the part of weight n, we therefore obtain:

$$s_n(b_1, \ldots, b_n) = \beta_1^n + \cdots + \beta_n^n.$$

The expression $\beta_1^n + \cdots + \beta_n^n$ is symmetric in the β_i, and can therefore be written as a polynomial in the b_j.

Examples:

$$\begin{aligned}
s_0 &= 1, & s_1 &= b_1, \\
s_2 &= b_1^2 - 2b_2, & s_3 &= 3b_3 - 3b_2 b_1 + b_1^3.
\end{aligned}$$

Now we give a lemma, which for a fixed power series Q displays all these values s_n as coefficients in a fixed power series. As usual, we write $f(x) = x/Q(x)$.

Lemma (Cauchy): *One has*

$$x\,\frac{f'(x)}{f(x)} = \sum_{j=0}^{\infty} (-1)^j s_j \cdot x^j.$$

Proof:

$$f(x) \ \equiv \ \frac{x}{(1 + \beta_1 x)\cdots(1 + \beta_n x)} \qquad (x^{n+2})$$

$$\Rightarrow \quad \frac{f'(x)}{f(x)} \ \equiv \ \frac{1}{x} - \frac{\beta_1}{1 + \beta_1 x} - \cdots - \frac{\beta_n}{1 + \beta_n x} \qquad (x^n)$$

$$\Rightarrow \quad x\,\frac{f'(x)}{f(x)} \ \equiv \ 1 - \frac{\beta_1 x}{1 + \beta_1 x} - \cdots - \frac{\beta_n x}{1 + \beta_n x} \qquad (x^{n+1})$$

$$\equiv \ 1 + \frac{-\beta_1 x}{1 - (-\beta_1 x)} + \cdots + \frac{-\beta_n x}{1 - (-\beta_n x)} \qquad (x^{n+1})$$

$$\equiv \ 1 + \sum_{j=1}^{\infty} (-\beta_1 x)^j + \cdots + \sum_{j=1}^{\infty} (-\beta_n x)^j \qquad (x^{n+1})$$

$$\equiv \ 1 + \sum_{j=1}^{\infty} (-1)^j \Big(\beta_1^j + \cdots + \beta_n^j\Big) \cdot x^j \qquad (x^{n+1}). \qquad \square$$

Example: $Q(x) = x/(1 - e^{-x})$, $f(x) = 1 - e^{-x}$. With these power series, one defines the Bernoulli numbers B_j by

$$Q(-x) = \sum_{j=0}^{\infty} \frac{B_j}{j!} x^j.$$

Then

$$x \frac{f'(x)}{f(x)} = x \frac{e^{-x}}{1 - e^{-x}} = x \frac{1}{e^x - 1} = \frac{-x}{1 - e^x} = Q(-x),$$

so that

$$s_j = (-1)^j \frac{B_j}{j!}.$$

For the Todd polynomials T_n, which one obtains as the multiplicative sequence for this power series Q, the coefficients s_n of c_n are therefore the same as the coefficients b_n in the power series Q. As one easily sees, these are also the coefficients of c_1^n.

Examples:

$$T_1 \;=\; \frac{c_1}{2}, \qquad T_2 \;=\; \frac{c_2 + c_1^2}{12},$$

$$T_3 \;=\; \frac{c_2 c_1}{24}, \qquad T_4 \;=\; \frac{-c_4 + c_3 c_1 + 3c_2^2 + 4c_2 c_1^2 - c_1^4}{720}.$$

If X is a manifold for which all Chern classes apart from the highest vanish, then its genus is given by

$$\varphi(X) = s_n \cdot c_n[X].$$

Later we shall encounter genera for which the power series $x \frac{f'(x)}{f(x)}$ has Eisenstein series as coefficients; the genus of such a manifold X would then be a multiple of the corresponding Eisenstein series.

One can also read off the s_n from the logarithm of the power series Q:

$$f(x) = \exp\left(\ln\left(f(x)\right)\right) = \exp\left(\int \frac{f'(x)}{f(x)}\, dx\right)$$

$$= \exp\left(\int \left(\frac{1}{x} \sum_{j=0}^{\infty} (-1)^j s_j \cdot x^j\right) dx\right)$$

$$= \exp\left(\ln(x) + \sum_{j=1}^{\infty} (-1)^j \frac{s_j}{j} x^j\right)$$

$$= x \cdot \exp\left(\sum_{j=1}^{\infty} (-1)^j \frac{s_j}{j} x^j\right)$$

$$\Rightarrow \quad Q(x) = \exp\left(\sum_{j=1}^{\infty} (-1)^{j+1} \frac{s_j}{j} x^j\right).$$

The s_n are therefore, up to the factor $(-1)^{n+1} \cdot \frac{1}{n}$, exactly the coefficients of the formal power series $\ln(Q(x))$.

2 Elliptic genera

2.1 The Weierstraß $\wp$-function

Let $\omega_1, \omega_2 \in \mathbb{C}$ such that $\tau := \omega_2/\omega_1 \notin \mathbb{R} \cup \{\infty\}$; we can number ω_1 and ω_2 such that $\mathrm{Im}\,(\tau) > 0$. Then $L = \mathbb{Z} \cdot \omega_2 + \mathbb{Z} \cdot \omega_1$ is a lattice in $\mathbb{C}$; put $L' = L \setminus \{0\}$. For $z \in \mathbb{C}$,

$$\wp(z) := \frac{1}{z^2} + \sum_{\omega \in L'} \left(\frac{1}{(z-\omega)^2} - \frac{1}{\omega^2} \right)$$

defines a meromorphic function with poles of order two at all lattice points. This function is called the Weierstraß $\wp$-function.

Definition: *A meromorphic function $f : \mathbb{C} \to \mathbb{C} \cup \{\infty\}$ is called **elliptic** (doubly periodic) with respect to a lattice $L \subset \mathbb{C}$ if $f(z+\omega) = f(z)$ for all $\omega \in L$.*

An elliptic function is therefore the same as a meromorphic function on the elliptic curve $E = \mathbb{C}/L$ which is isomorphic to a torus.

Examples: The $\wp$-function is an even function. Since

$$\wp'(z) = \sum_{\omega \in L} \frac{-2}{(z-\omega)^3}$$

is an odd elliptic function, one has $\wp(z + \omega_i) - \wp(z) \equiv c_i$; putting $z = -(\omega_i/2)$ shows that $c_i = 0$. Therefore the $\wp$-function is elliptic. Moreover, $\wp$ satisfies the following differential equation:

$$\wp'(z)^2 = 4 \cdot \wp(z)^3 - g_2 \cdot \wp(z) - g_3,$$

where g_2 and g_3 depend on the lattice.

With the definition $s_r(L) := \sum_{\omega \in L'} \frac{1}{\omega^r}$ we obtain series which are absolutely convergent for $r > 2$; we have $s_r(L) = 0$ for odd r (since for $\omega \in L$ also $-\omega \in L$). Moreover: $g_2(L) = 60 \cdot s_4(L)$, $g_3(L) = 140 \cdot s_6(L)$.

For elliptic functions there are theorems of Liouville-type (cf. [La73]):

Theorem: *For $f(z)$ elliptic with respect to L, the following hold on $\mathbb{C}/L$:*

1) *The number of zeroes of f is equal to the number of poles (counted with multiplicities!).*

2) *The sum of the residues is zero.*

3) *Let $(f) = \sum_{a \in \mathbb{C}/L} n_a \cdot (a)$ with $n_a \in \mathbb{Z}$ be the divisor of f, then $\sum_{a \in \mathbb{C}/L} n_a \cdot a \equiv 0$ modulo L, i.e. $\sum_{a \in \mathbb{C}/L} n_a \cdot a \in L$.* ⬜

Corollary: *There is no elliptic function which has precisely one pole of order one.*

The elliptic function $\wp'$ has a pole of order three in $\mathbb{C}/L$ and therefore also three zeroes. Since $\wp'$ is odd, it follows from $\wp'(-z) = -\wp'(z)$ that for those z with $-z \equiv z\,(L)$, either $\wp'(z) = 0$ or $\wp$ has a pole at z. However, $-z \equiv z\,(L) \Leftrightarrow 2 \cdot z \in L$, and on the torus $\mathbb{C}/L \cong S^1 \times S^1$ there are exactly three non-trivial two-division points; namely, for the chosen basis ω_1, ω_2 of the lattice L these are the points $\omega_1/2$, $\omega_2/2$ and $(\omega_1 + \omega_2)/2$. At these points the $\wp$-function has no pole, so the function $\wp'$ vanishes there.

A standard notation is now:

$$e_1 := \wp\Big(\frac{\omega_1}{2}\Big),\ e_2 := \wp\Big(\frac{\omega_2}{2}\Big),\ e_3 := \wp\Big(\frac{\omega_1 + \omega_2}{2}\Big).$$

In terms of e_1, e_2 and e_3 the differential equation for $\wp$ can now also be written as

$$\begin{aligned}
\wp'(z)^2 &= 4\wp(z)^3 - g_2 \cdot \wp(z) - g_3 \\
&= 4(\wp(z) - e_1)(\wp(z) - e_2)(\wp(z) - e_3).
\end{aligned}$$

One therefore has:

$$\begin{aligned}
e_1 + e_2 + e_3 &= 0, \\
4(e_1 e_2 + e_1 e_3 + e_2 e_3) &= -g_2, \\
4 e_1 e_2 e_3 &= g_3.
\end{aligned}$$

The $\wp$-function takes each value exactly twice; since it is an even function, four times the two preimages coincide, namely exactly for the two-division points. For the other values the two preimages are distinguished by the sign of $\wp'(z)$. One therefore obtains by means of $\wp$ a double covering of $P_1(\mathbb{C})$ by the torus $\mathbb{C}/L$, which is branched at the two-division points. The cubic curve in $\mathbb{C}^2 \subset P_2(\mathbb{C})$ given by the equation $y^2 = 4x^3 - g_2 x - g_3$ is smooth and has exactly one point on the line at infinity in $P_2(\mathbb{C})$. On the basis of the differential equation for $\wp$, it is parametrized by $x = \wp(z)$, $y = \wp'(z)$, where the point $0 \in \mathbb{C}/L$ is mapped to the point of the curve lying at infinity (inflection point). By means of this isomorphism the cubic curve obtains a group structure:

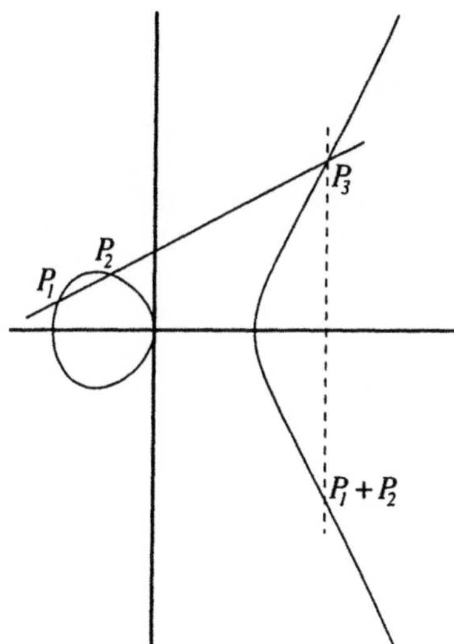

Group structure of a cubic curve

Proposition: *Three points P_1, P_2 and P_3 on the curve lie on a line, if and only if $P_1 + P_2 + P_3 = 0$.*

Proof: The linear equation $a \cdot x + b \cdot y + c = 0$ restricted to the curve yields, in view of the parametrization by $\wp$ and $\wp'$, the equation

$$h(z) := a \cdot \wp(z) + b \cdot \wp'(z) + c = 0.$$

Now h is an elliptic function, i.e. a meromorphic function on $\mathbb{C}/L$ with a pole of order three at $0 \in \mathbb{C}/L$, hence it has also exactly three zeroes z_1, z_2, z_3. For these there holds according to Liouville:

$$z_1 + z_2 + z_3 - 3 \cdot 0 = 0 \in \mathbb{C}/L.$$

Because we require that the $\wp$-function be a group isomorphism of $\mathbb{C}/L$ onto the cubic curve, the sum of the three intersection points of the curve with the line must likewise be zero.

Remark: By explicitly calculating the intersection points of the cubic with the line, one obtains the addition theorem for the $\wp$-function.

2.2 Construction of elliptic genera

The $\wp$-function is an even function and takes the values e_1, e_2 and e_3 with order two, i.e. $\wp(z) - e_1$ has a double zero at $\omega_1/2$. Therefore we can take the square root, and

$f(z) := 1/\sqrt{\wp(z) - e_1}$ with the choice of sign $\sqrt{\wp(z) - e_1} = \frac{1}{z} + \cdots$ is defined and single-valued on $\mathbb{C}$. The odd function f is meromorphic on $\mathbb{C}$, with zeroes of order 1 at all lattice points and poles of order 1 at $(\omega_1/2) + L$. Although f is not elliptic with respect to L, there exists a sublattice $\widetilde{L}$ of index 2 in L with respect to which f is elliptic. The projection $\mathbb{C}/\widetilde{L} \to \mathbb{C}/L$ is an unbranched double covering. The divisor of f is

$$(f) = 0 + \alpha - (\omega_1/2) - (\omega_1/2 + \alpha).$$

Because $(f) \in \widetilde{L}$ (cf. part 3 of the theorem in section 2.1), we have $\omega_1 \in \widetilde{L}$; therefore $\widetilde{L} = \mathbb{Z} \cdot 2\omega_2 + \mathbb{Z} \cdot \omega_1$ and $f(z + \omega_2) = -f(z)$.

Theorem: *The genus defined by the function $f(z) = 1/\sqrt{\wp(z) - e_1}$ is an elliptic genus as defined in section 1.7, i.e. for the derivative g' of the inverse function and the duplication formula hold:*

$$g'(y) = 1 + \delta \cdot y^2 + \cdots \qquad and \qquad f(2x) = \frac{2f(x)f'(x)}{1 - \varepsilon \cdot f(x)^4}.$$

Further, f satisfies the differential equation

$$f'(z)^2 = 1 - 2\delta \cdot f(z)^2 + \varepsilon \cdot f(z)^4,$$

and has the addition theorem

$$f(u + v) = \frac{f(u)f'(v) + f'(u)f(v)}{1 - \varepsilon \cdot f(u)^2 f(v)^2}.$$

Here $\delta = -\frac{3}{2} \cdot e_1$, and $\varepsilon = (e_1 - e_2)(e_1 - e_3)$.

Remark: Since the $\wp$-function assumes each value twice, counted with multiplicity, the e_j as values at the two-division points are distinct, and therefore $\varepsilon = (e_1 - e_2)(e_1 - e_3) \neq 0$ and $\delta^2 - \varepsilon = \frac{1}{4}(e_2 - e_3)^2 \neq 0$. So we cannot obtain the $\hat{A}$- and L-genus in this way, because there we have $\varepsilon = 0$ resp. $\delta^2 - \varepsilon = 0$.

Proof of the theorem: We consider the curve in $P_2(\mathbb{C})$ which is parametrized by $\xi = f(z)$ and $\eta = f'(z)$. With $x = \wp(z)$ and $y = \wp'(z)$ one has:

$$\xi = \frac{1}{\sqrt{\wp(z) - e_1}} = \frac{1}{\sqrt{x - e_1}}$$
$$\Rightarrow \quad x = \xi^{-2} + e_1,$$
$$\eta = f'(z) = \left((\wp(z) - e_1)^{-1/2}\right)' = -\frac{\wp'(z)}{2} \cdot (\wp(z) - e_1)^{-3/2}$$
$$= -\frac{y}{2} \cdot (x - e_1)^{-3/2} = -\frac{y}{2} \cdot \xi^3$$
$$\Rightarrow \quad y = -2\eta\xi^{-3}.$$

The (ξ, η)-curve satisfies the following equation:

$$y^2 = 4(x - e_1)(x - e_2)(x - e_3) = 4x^3 - g_2 x - g_3$$
$$\Rightarrow \quad 4\eta^2 \xi^{-6} = 4\xi^{-2}\left(\xi^{-2} + e_1 - e_2\right)\left(\xi^{-2} + e_1 - e_3\right)$$
$$\Rightarrow \quad \eta^2 = \left(1 + \xi^2(e_1 - e_2)\right)\left(1 + \xi^2(e_1 - e_3)\right)$$
$$= 1 + (3e_1 - e_1 - e_2 - e_3)\xi^2 + (e_1 - e_2)(e_1 - e_3)\xi^4$$
$$= 1 + 3e_1\xi^2 + (e_1 - e_2)(e_1 - e_3)\xi^4$$
$$= 1 - 2\delta \cdot \xi^2 + \varepsilon \cdot \xi^4,$$

with $\delta = -\frac{3}{2} \cdot e_1$ and $\varepsilon = (e_1 - e_2)(e_1 - e_3)$. Therefore $f(z) = (\wp(z) - e_1)^{-1/2}$ indeed satisfies the differential equation

$$f'(z)^2 = 1 - 2\delta \cdot f(z)^2 + \varepsilon \cdot f(z)^4$$

for the given values of δ and ε. The (ξ, η)-curve is obtained from the elliptic curve $\mathbb{C}/\widetilde{L}$ by the identification of $(\omega_1/2)$ with $(\omega_1/2) + \omega_2$. From this there arises a singularity of type A_3 at infinity. The formulas for x and y yield a covering of degree two of the (x, y)-curve by the (ξ, η)-curve.

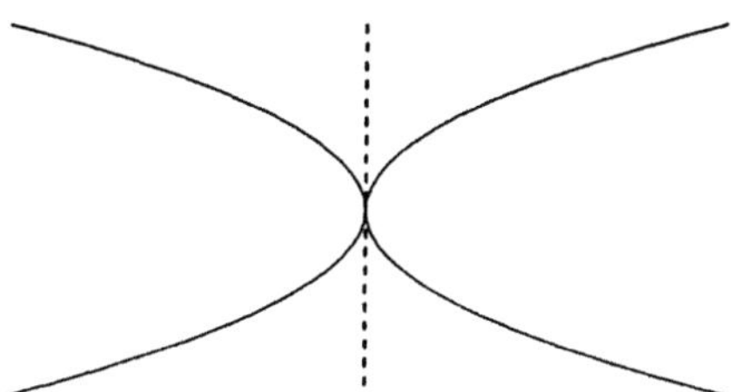

Singularity of type A_3

The derivation of the addition theorem for f was already carried out by Fagnano in 1718 for duplication in connection with an investigation on the lemniscate, and was given in general by Euler (cf. [Eu251]): put

$$R(u, v) := \frac{f(u)f'(v) + f'(u)f(v)}{1 - \varepsilon \cdot f(u)^2 f(v)^2}$$

and consider the total differential

$$dR(u, v) = \frac{\partial R}{\partial u}\, du + \frac{\partial R}{\partial v}\, dv.$$

We have

$$(\star) \quad \frac{\partial R}{\partial u} \cdot \left(1 - \varepsilon f(u)^2 f(v)^2\right)^2 = \left(f'(u)f'(v) + f''(u)f(v)\right) \cdot \left(1 - \varepsilon f(u)^2 f(v)^2\right)$$
$$-\left(f(u)f'(v) + f'(u)f(v)\right) \cdot \left(-2\varepsilon f(u)f'(u)f(v)^2\right).$$

Now we calculate f'' from

$$f'(u)^2 = 1 - 2\delta f(u)^2 + \varepsilon f(u)^4$$
$$\Rightarrow \quad 2f'(u)f''(u) = -4\delta f(u)f'(u) + 4\varepsilon f(u)^3 f'(u)$$
$$\Rightarrow \quad f''(u) = -2\delta f(u) + 2\varepsilon f(u)^3.$$

We now substitute this into $(\bigstar)$ and then calculate the non-symmetric component in u and v (S denotes the additive subgroup of symmetric functions in u and v):

$$\frac{\partial R}{\partial u} \cdot \left(1 - \varepsilon f(u)^2 f(v)^2\right)^2 \equiv 2\varepsilon f(u)^3 f(v) \cdot \left(1 - \varepsilon f(u)^2 f(v)^2\right)$$

$$+ 2\varepsilon f(u) f(v)^3 f'(u)^2 \tag{S}$$

$$\equiv 2\varepsilon f(u)^3 f(v) - 2\varepsilon^2 f(u)^5 f(v)^3$$

$$+ 2\varepsilon f(u) f(v)^3 f'(u)^2 \tag{S}$$

$$\equiv 2\varepsilon f(u)^3 f(v) - 2\varepsilon^2 f(u)^5 f(v)^3$$

$$+ 2\varepsilon f(u) f(v)^3 \cdot \left(1 - 2\delta f(u)^2 + \varepsilon f(u)^4\right) \tag{S}$$

$$\equiv 0 \tag{S}$$

Therefore $\frac{\partial R}{\partial u} = s(u,v)$ for some symmetric functions in u and v, hence so is $\frac{\partial R}{\partial v} = s(v,u) = s(u,v)$, and $dR(u,v) = s(u,v) \cdot (du + dv) = s(u,v) \cdot d(u+v)$. For $u + v = c$ constant it now follows from $dR(u,v) = 0$ that $R(u,v) = \tilde{c}$ is constant, with $\tilde{c} = R(0,c) = f(c) = f(u+v)$. This proves the addition formula and gives the duplication formula as a special case:

$$f(2z) = \frac{2f(z)f'(z)}{1 - \varepsilon \cdot f(z)^4} \, .$$

Remark: If one lets the lattice degenerate by $\omega_1 \to 0$, then one obtains with $e_2 = e_3$ the case $\varepsilon = \delta^2$. Through this limit process, the function f for the degenerate lattice must satisfy the differential equation $f' = 1 - \delta \cdot f^2$, hence can be written in the form $f(x) = \tanh(\alpha x)/\alpha$ where α is determined by δ. For $\delta = 1$ (hence $\alpha = 1$) we obtain the L-genus.

In the other two degenerate cases $\omega_1 \to \omega_2$ ($e_1 = e_2$), resp. $\omega_2 \to 0$ ($e_1 = e_3$) one has $\varepsilon = 0$. The differential equation becomes $f'^2 = 1 - 2\delta \cdot f^2$ and has the solutions $f(x) = \sinh(\alpha x)/\alpha$. In particular, for $\delta = -\frac{1}{8}$ ($\alpha = \frac{1}{2}$) one obtains the $\hat{A}$-genus.

The above construction yields for each lattice $L \subset \mathbb{C}$ an elliptic genus, where δ and ε depend on L. On the other hand, if δ and ε are given then e_1, e_2 and e_3 can be determined (notice that $e_1 + e_2 + e_3 = 0$). The discriminant of the equation $y^2 = 4(x - e_1)(x - e_2)(x - e_3)$ is

$$\Delta = 64\varepsilon^2\left(\delta^2 - \varepsilon\right).$$

If Δ is different from zero, then the equation determines an elliptic curve $\mathbb{C}/L$, where $L \subset \mathbb{C}$ is a lattice, which yields the elliptic genus corresponding to δ and ε by means of the above construction. If the discriminant Δ vanishes, then either $\varepsilon = 0$ or $\varepsilon = \delta^2$ and the corresponding elliptic genus arises by degeneration of the lattice L.

In view of the differential equation which is satisfied by the power series f for elliptic genera, it is clear that the coefficient of x^{2k} in $Q(x) = x/f(x)$ is a homogeneous polynomial of weight $2k$ in δ and ε, to which we assign the weights 2 resp. 4. Hence the elliptic genus of a compact, oriented, differentiable manifold M^{4k} is a homogeneous polynomial of weight $2k$ in δ and ε. Moreover, δ and ε are lattice invariants of weights 2 and 4, i.e. for each lattice $L \subset \mathbb{C}$ and $\lambda \in \mathbb{C}$, $\lambda \neq 0$, we have $\delta(\lambda \cdot L) = \lambda^{-2} \cdot \delta(L)$, resp. $\varepsilon(\lambda \cdot L) = \lambda^{-4} \cdot \varepsilon(L)$. Hence the elliptic genus of a manifold M^{4k} is a lattice invariant of weight $2k$. Further lattice invariants of weight $2k$ are the earlier defined $s_{2k} = \sum_{\omega \in L'} \frac{1}{\omega^{2k}}$. The lattice invariants δ and ε are, as we have seen, really lattice invariants for lattices with distinguished two-division points. For the precise connection between elliptic genera and modular forms on the congruence subgroup $\Gamma_0(2)$ of $\mathrm{SL}_2(\mathbb{Z})$ we refer to Appendix I.

As was mentioned earlier, an elliptic genus is already determined by its values on $P_2(\mathbb{C})$ and $P_k(\mathbb{H})$. Indeed, we have (cf. section 1.7):

$$\varphi(P_2(\mathbb{C})) = \delta,$$

$$\varphi(P_k(\mathbb{H})) = \begin{cases} \varepsilon^{k/2}, & k \equiv 0 \ (2), \\ 0, & k \equiv 1 \ (2). \end{cases}$$

2.3 An excursion on the lemniscate

The lemniscate is the geometric locus of all points of the Euclidean plane, for which the product of the distances to the foci $(a,0)$ and $(-a,0)$ is equal to a^2.

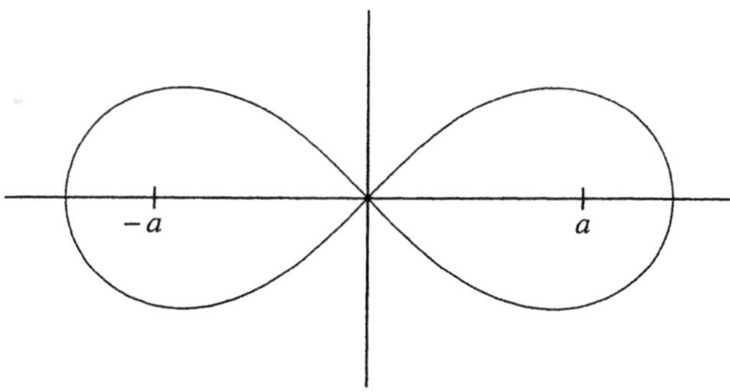

The lemniscate

We have therefore the defining equation

$$\left((x-a)^2 + y^2\right) \cdot \left((x+a)^2 + y^2\right) = a^4.$$

We introduce the new coordinate $r^2 = x^2 + y^2$ and obtain

$$\left(r^2 - 2ax + a^2\right) \cdot \left(r^2 + 2ax + a^2\right) = a^4$$
$$\Rightarrow \quad \left(r^2 + a^2\right)^2 - 4a^2 x^2 = a^4$$
$$\Rightarrow \quad r^4 + 2a^2 r^2 - 4a^2 x^2 = 0.$$

Without restriction, normalize a so that $2a^2 = 1$:

$$\Rightarrow \quad r^4 + r^2 - 2x^2 = 0$$
$$\Rightarrow \quad r^4 + r^2 - 2\left(r^2 - y^2\right) = 0 \qquad (\text{since } x^2 + y^2 = r^2)$$
$$\Rightarrow \quad r^4 - r^2 + 2y^2 = 0.$$

Hence:

$$2x^2 = r^2 + r^4 \qquad \text{and} \qquad 2y^2 = r^2 - r^4.$$

This yields a parametrization of the lemniscate by the parameter r. In the following, $\dot{x}$ and $\dot{y}$ denote differentiation with respect to r.

For the arc length of the lemniscate we now obtain

$$2x\dot{x} = r + 2r^3 \qquad \text{and} \qquad 2y\dot{y} = r - 2r^3$$
$$\Rightarrow \quad \dot{x}^2 = \frac{\left(r + 2r^3\right)^2}{4x^2} = \frac{r^2\left(1 + 2r^2\right)^2}{2(r^2 + r^4)} = \frac{\left(1 + 2r^2\right)^2}{2(1 + r^2)} .$$

Analogously:

$$\dot{y}^2 = \frac{\left(1 - 2r^2\right)^2}{2(1 - r^2)} .$$

Hence we obtain

$$\dot{x}^2 + \dot{y}^2 = \frac{\left(1 + 2r^2\right)^2\left(1 - r^2\right) + \left(1 - 2r^2\right)^2\left(1 + r^2\right)}{2(1 - r^4)} = \frac{1}{1 - r^4},$$

and therefore

$$\text{arc length} = \int_0^r \sqrt{\dot{x}^2 + \dot{y}^2}\, dt = \int_0^r \frac{dt}{\sqrt{1 - t^4}} .$$

For the elliptic genus we have obtained

$$g(y) = \int_0^y g'(t)\, dt = \int_0^y \frac{dt}{f'(g(t))}$$

$$= \int_0^y \frac{dt}{\sqrt{1 - 2\delta f(g(t))^2 + \varepsilon f(g(t))^4}}$$

$$= \int_0^y \frac{dt}{\sqrt{1 - 2\delta \cdot t^2 + \varepsilon \cdot t^4}}\,.$$

The arc length of the lemniscate therefore corresponds to the logarithm g of an elliptic genus for the special case $\delta = 0$ and $\varepsilon = -1$.

A classical problem, which as was mentioned above was already treated by Fagnano in the year 1718, concerns the duplication of the arc: For a given parameter r, find the parameter R corresponding to the doubled arc length.

Put

$$P(t) := 1 - 2\delta \cdot t^2 + \varepsilon \cdot t^4 \qquad \text{and} \qquad x_i := g(y_i) = \int_0^{y_i} \frac{dt}{\sqrt{P(t)}}$$

for $i = 1, 2$ and $y_i \in \mathbb{C}$.

As we have previously seen, with $y_i = f(x_i)$ we have

$$x_1 + x_2 = g(y_1) + g(y_2) = g(f(x_1 + x_2))$$

$$= g\left(\frac{f(x_1)f'(x_2) + f'(x_1)f(x_2)}{1 - \varepsilon f(x_1)^2 f(x_2)^2} \right)$$

$$= g\left(\frac{y_1 \sqrt{P(y_2)} + y_2 \sqrt{P(y_1)}}{1 - \varepsilon y_1^2 y_2^2} \right) = g(y_1 + y_2 + \cdots).$$

For the lemniscate this means:

$$R = \frac{2r \cdot \sqrt{1 - r^4}}{1 + r^4} = 2r + \cdots.$$

For Fagnano, it was an interesting question whether the parameter R could be constructed from r by means of circles and lines (i.e. "ruler and compass"). The formula above uses only addition, square root and product; therefore R is indeed geometrically constructible.

2.4 Geometric complement on the addition theorem

As we have seen, the elliptic functions f and f' associated to a lattice $\widetilde{L}$ parametrize a curve in $P_2(\mathbb{C})$ which is given by $y^2 = 1 - 2\delta \cdot x^2 + \varepsilon \cdot x^4$. The group law on the elliptic curve $\mathbb{C}/\widetilde{L}$ thereby corresponds to the addition theorem for the elliptic genus. J. Nekovář has pointed out a beautiful geometric proof of the group law in [Ma79].

We therefore consider the curve C with equation $y^2 z^2 = z^4 - 2\delta \cdot x^2 z^2 + \varepsilon \cdot x^4$ in homogeneous coordinates. It has a unique singular point, namely $P = (0 : 1 : 0)$; and the line $\{z = 0\}$ is doubly tangent at P, touching the curve there to order four. Under the parametrization of the curve by f and f', the points $(\omega_1/2)$ and $(\omega_1/2) + \omega_2$ in $\mathbb{C}/\widetilde{L}$ are both mapped to P. Away from the singularity, the map is one-to-one. The neutral element 0 of the group $\mathbb{C}/\widetilde{L}$ is mapped to the point $Q = (0 : 1 : 1)$. Now let $yz = z^2 + A \cdot xz + B \cdot x^2$ be a family of parabolas. To any two points on C there is a parabola in the family which goes through both. This cuts our curve in 8 points; among these, 4 intersection points lie in the affine part (including Q) while the point P is always an intersection point of order four. Let (x_1, y_1), (x_2, y_2) and (x_3, y_3) be the three intersection points in the affine part different from the neutral element.

The meromorphic function $h := \left(yz - z^2 - Axz - Bx^2\right)/z^2$ on the curve has as divisor

$$(h) = (0 : 1 : 1) + (x_1 : y_1 : 1) + (x_2 : y_2 : 1) + (x_3 : y_3 : 1) + 4 \cdot P - 8 \cdot P.$$

If we pull h back to the elliptic curve by means of the parametrization $(x : y : z) = (f(t) : f'(t) : 1)$, then the divisor there is:

$$
\begin{aligned}
\left(\widetilde{h}\right) &= (0) + (t_1) + (t_2) + (t_3) - 2 \cdot (\omega_1/2) - 2 \cdot (\omega_1/2 + \omega_2) \\
\Rightarrow \qquad & t_1 + t_2 + t_3 \equiv 0 \qquad \left(\widetilde{L}\right) \\
\Rightarrow \qquad & \qquad -t_3 \equiv t_1 + t_2 \; \left(\widetilde{L}\right) \\
\Rightarrow \qquad & -x_3 = -f(t_3) = f(-t_3) = f(t_1 + t_2).
\end{aligned}
$$

We have to express $-x_3 = f(t_1 + t_2)$ in terms of $x_1 = f(t_1)$ and $x_2 = f(t_2)$, to derive the addition theorem for f.

To do this, substitute the equation of the parabolas into the equation of the curve (now affine again):

$$
\begin{aligned}
& \left(B^2 - \varepsilon\right) \cdot x^4 + 2AB \cdot x^3 + \left(A^2 + 2B + 2\delta\right) \cdot x^2 + 2A \cdot x = 0 \\
\Rightarrow \; & \left(B^2 - \varepsilon\right) \cdot x^3 + 2AB \cdot x^2 + \left(A^2 + 2B + 2\delta\right) \cdot x + 2A = 0 \\
\Rightarrow \; & x_1 + x_2 + x_3 = \frac{2AB}{\varepsilon - B^2} \quad \text{and} \quad x_1 x_2 x_3 = \frac{2A}{\varepsilon - B^2} \\
\Rightarrow \; & x_1 + x_2 + x_3 = B \cdot x_1 x_2 x_3 \quad \text{so} \quad x_3 = \frac{x_1 + x_2}{B \cdot x_1 x_2 - 1},
\end{aligned}
$$

since x_1, x_2, x_3 are precisely the roots of the polynomial.

$$y_1 = 1 + A \cdot x_1 + B \cdot x_1^2, \quad y_2 = 1 + A \cdot x_2 + B \cdot x_2^2$$

$$\Rightarrow \quad x_1 y_2 - x_2 y_1 = (x_1 - x_2)(1 - B \cdot x_1 x_2) \tag{I},$$

$$y_1^2 = 1 - 2\delta \cdot x_1^2 + \varepsilon \cdot x_1^4, \quad y_2^2 = 1 - 2\delta \cdot x_2^2 + \varepsilon \cdot x_2^4$$

$$\Rightarrow \quad x_1^2 y_2^2 - x_2^2 y_1^2 = \left(x_1^2 - x_2^2\right)\left(1 - \varepsilon \cdot x_1^2 x_2^2\right) \tag{II},$$

$$(\text{I}) \;\&\; (\text{II}) \quad \Rightarrow \quad x_1 y_2 + x_2 y_1 = \frac{x_1^2 y_2^2 - x_2^2 y_1^2}{x_1 y_2 - x_2 y_1}$$

$$= (x_1 + x_2) \frac{1 - \varepsilon \cdot x_1^2 x_2^2}{1 - B \cdot x_1 x_2}$$

$$= -x_3 \cdot \left(1 - \varepsilon \cdot x_1^2 x_2^2\right)$$

$$\Rightarrow \quad -x_3 = \frac{x_1 y_2 + x_2 y_1}{1 - \varepsilon \cdot x_1^2 x_2^2}.$$

Together with the parametrization $x_i = f(t_i)$, $y_i = f'(t_i)$ we obtain

$$-x_3 = f(t_1 + t_2) = \frac{f(t_1) f'(t_2) + f(t_2) f'(t_1)}{1 - \varepsilon f(t_1)^2 f(t_2)^2}.$$

3 A universal addition theorem for genera

3.1 Virtual submanifolds

Let M^{2n} be a compact, oriented, differentiable manifold and $U \subset M$ an oriented, differentiable submanifold of codimension two. Let $u \in H^2(M; \mathbb{Z})$ be the Poincaré-dual cohomology class of the fundamental cycle of U. Further, let TU and TM be the tangent bundles of U, resp. M, and let NU be the normal bundle of U in M. By choosing a Riemannian metric we get an isomorphism $TU \oplus NU \cong TM|_U$ and NU gets the structure of an $O(2)$-bundle. Since M and U are oriented, the structure group of NU can be further reduced to $SO(2) \cong U(1)$. The bundle NU can therefore be considered as a complex line bundle which has the first Chern class $i^*(u)$ ($i: U \hookrightarrow M$ inclusion), i.e.:

$$c(NU) = 1 + c_1(NU) = 1 + i^*(u)$$
$$\Rightarrow \quad p(NU) = 1 + i^*(u^2).$$

Using $TU \oplus NU = i^*(TM)$, we obtain

$$p(U) \cdot p(NU) = i^* p(M)$$
$$\Rightarrow \quad p(U) \cdot i^*(1 + u^2) = i^* p(M)$$
$$\Rightarrow \quad p(U) = i^*\left(p(M) \cdot (1 + u^2)^{-1}\right)$$
$$= i^*\left((1 + p_1(M) + p_2(M) + \cdots) \cdot (1 - u^2 + u^4 - u^6 \pm \cdots)\right).$$

Example:

$$p_1(U) = i^*\left(p_1(M) - u^2\right),$$
$$p_2(U) = i^*\left(p_2(M) - p_1(M) \cdot u^2 + u^4\right).$$

From topology one knows that for $a \in H^*(M; \mathbb{Z})$ holds:

$$i^*(a)[U] = (a \cdot u)[M].$$

Therefore the Pontrjagin numbers of U are obtained by evaluation of a homogeneous polynomial of degree $2n$ in u and the $p_i(M)$ on the fundamental cycle of the manifold M. The equivalence class $[U]$ in the cobordism ring $\Omega \otimes \mathbb{Q}$ therefore depends, for given M, only on the cohomology class u.

According to Thom [Th54], in our situation each cohomology class of dimension 2 can be represented by means of Poincaré duality by a submanifold of codimension 2; therefore $u \in H^2(M; \mathbb{Z})$ is called a virtual submanifold. Let $u_1, \ldots, u_r \in H^2(M; \mathbb{Z})$; then

$(u_1, \ldots, u_r)$ is called a virtual submanifold of codimension $2r$. This is represented by a manifold in $\Omega^{2n-2r} \otimes \mathbb{Q}$, which one obtains as follows: Represent u_1 by a submanifold U_1 of M, represent the restriction of u_2 to U_1 by a submanifold U_2 of U_1, etc., obtaining further submanifolds $U_3, \ldots, U_r$. The manifold U_r is then a submanifold of M of codimension $2r$, which represents $(u_1, \ldots, u_r)$. An alternative construction is the following: If $u_1, \ldots, u_r$ can be represented by submanifolds $U_1', \ldots, U_r'$ of M, which are in general position (transversal to one another), then their intersection represents $(u_1, \ldots, u_r)$.

Example: For $M = P_n(\mathbb{C})$, the $u_1, \ldots, u_r$ are integral multiples $u_i = d_i \cdot g$ of the generator g of $H^2(P_n(\mathbb{C}); \mathbb{Z}) \cong \mathbb{Z}$, and correspond exactly to the hypersurfaces of degree d_i. The virtual submanifolds $(u_1, \ldots, u_r)$ correspond to complete intersections.

Analogous to the above, one obtains:

$$p(U_r) = i^*\big(p(M) \cdot (1 + u_1^2)^{-1} \cdots (1 + u_r^2)^{-1}\big)$$
$$= i^*(1 + \widetilde{p}_1 + \widetilde{p}_2 + \cdots).$$

The Pontrjagin numbers of U_r are therefore again polynomials in $p_i(M)$ and the u_j evaluated on M, since for $4 \cdot \sum_{j=1}^{k} i_j = 2n - 2r = \dim U_r$ we have:

$$(p_{i_1}(U_r) \cdots p_{i_k}(U_r))[U_r] = i^*(\widetilde{p}_{i_1}(U_r) \cdots \widetilde{p}_{i_k}(U_r))[U_r]$$
$$= (\widetilde{p}_{i_1}(U_r) \cdots \widetilde{p}_{i_k}(U_r) \cdot u_1 \cdots u_r)[M].$$

Now let $Q(x)$, as usual, be an even and $f(x)$, $g(y)$ be odd power series with $f(x) = x/Q(x)$, $f(x) = y$, $g(y) = x$. Then the multiplicative sequence of polynomials $K_n(p_1, \ldots, p_n)$ associated to the power series $Q(x)$ was defined by

$$Q(x_1) \cdots Q(x_k) = \sum_{n=0}^{k} K_n(p_1, \ldots, p_n) + \sum_{n=k+1}^{\infty} K_n(p_1, \ldots, p_k, 0, \ldots, 0)$$

(p_i is the i-th elementary symmetric function in the x_j^2). Further, let φ be the genus given by this multiplicative sequence, i.e.:

$$\varphi(M) := \Big(\sum_{n=0}^{\infty} K_n(p_1, \ldots, p_n)\Big)[M] = K(TM)[M].$$

In our situation this means:

$$\varphi(U_r) = i^*\Big(K(TM)\frac{f(u_1) \cdots f(u_r)}{u_1 \cdots u_r}\Big)[U_r]$$
$$= (K(TM) \cdot f(u_1) \cdots f(u_r))[M].$$

Problem: Can the genus φ of a submanifold represented by $u + v$ ($u, v \in H^2(M; \mathbb{Z})$) be calculated from the genera of the virtual submanifolds $(\underbrace{u, \ldots, u}_{r}, \underbrace{v, \ldots, v}_{s})$?

Since the power series $f(x)$ begins with x, the power series $f(u+v)$ can be formally expressed as a power series in $f(u)$ and $f(v)$. With $y_1 := f(u)$, $y_2 := f(v)$ we therefore have:

$$
\begin{aligned}
f(u+v) &= f(g(f(u)) + g(f(v))) \\
&= f(g(y_1) + g(y_2)) \\
&= \sum_{r,s \geqslant 0} a_{rs} \cdot y_1^r y_2^s \\
&= F(y_1, y_2) = y_1 + y_2 + \cdots.
\end{aligned}
$$

Here g is the inverse function of f. The power series F is called the formal group law of f. From the definition of F it follows that

$$
g(F(y_1, y_2)) = g(y_1) + g(y_2).
$$

Therefore g is also called the logarithm of the formal group law F. For an elliptic genus, we have already computed as an example (cf. section 2.2):

$$
F(y_1, y_2) = \frac{y_1 \cdot \sqrt{P(y_2)} + y_2 \cdot \sqrt{P(y_1)}}{1 - \varepsilon \cdot y_1^2 y_2^2},
$$

where $P(y) = 1 - 2\delta \cdot y^2 + \varepsilon \cdot y^4$. Now let u, $v \in H^2(M; \mathbb{Z})$ be given, then we have the following

Lemma: *Let a_{rs} be the coefficients of the formal group law F of the power series f. Then*

$$
\varphi((u+v)) = \sum_{r,s \geqslant 0} a_{rs} \cdot \varphi\big(\big(\underbrace{u, \ldots, u}_{r}, \underbrace{v, \ldots, v}_{s}\big)\big).
$$

Remark: Note that the coefficients a_{rs} are independent of u, v and M.

Proof of the lemma:

$$
\begin{aligned}
\varphi((u+v)) &= (K(TM) \cdot f(u+v))[M] \\
&= \Big(K(TM) \cdot \sum_{r,s \geqslant 0} a_{rs} \cdot f(u)^r f(v)^s\Big)[M] \\
&= \sum_{r,s \geqslant 0} a_{rs} \cdot \varphi((\underbrace{u, \ldots, u}_{r}, \underbrace{v, \ldots, v}_{s})). \qquad \square
\end{aligned}
$$

Example: For a 6-dimensional manifold M^6 and the elliptic genus this means

$$
\begin{aligned}
\varphi(u+v) &= \varphi(u) + \varphi(v) - \delta \cdot \varphi(u, u, v) - \delta \cdot \varphi(v, v, u) \\
&= \varphi(u) + \varphi(v) - \delta \cdot (\varphi(u, u, v) + \varphi(v, v, u)) \\
&= \varphi(u) + \varphi(v) - \delta \cdot \big(u^2 v[M] + v^2 u[M]\big),
\end{aligned}
$$

since the virtual submanifold of M which corresponds to (u, u, v), resp. (v, v, u) has dimension zero (points).

Since $\Omega^4 \otimes \mathbb{Q}$ is only one-dimensional, there is (up to normalization by means of δ) also only one genus here, namely for $\delta = 1$ the L-genus. Therefore

$$L(u + v) = L(u) + L(v) - u^2 v[M] - v^2 u[M].$$

3.2 A universal genus

The considerations about formal group laws can now be carried out for a universal genus (cf. [Oc87] and [Bu70]).

Take as genus the homomorphism $\underline{\varphi} = \mathrm{id}$ of the cobordism ring into itself. It is now interesting to ask about the characteristic power series and especially about the addition formula for this universal genus. We have

$$\underline{\varphi} = \mathrm{id} : \Omega \otimes \mathbb{Q} \to \Omega \otimes \mathbb{Q},$$

$$\varphi((u + v)) = \sum_{r,s} \underline{a}_{rs} \cdot \underline{\varphi}((\underbrace{u, \ldots, u}_{r}, \underbrace{v, \ldots, v}_{s}))$$

$$\Rightarrow \quad (u + v) = \sum_{r,s} \underline{a}_{rs} \cdot (\underbrace{u, \ldots, u}_{r}, \underbrace{v, \ldots, v}_{s}),$$

$$(\mathrm{codim}\ 2) \qquad\qquad (\mathrm{codim}\ 2(r + s))$$

with cobordism classes $\underline{a}_{rs} \in \Omega \otimes \mathbb{Q}$ of dimension $2(r + s - 1)$.

Remark: The above equation is an identity in the cobordism ring. Hence it is respected by all genera (ring homomorphisms). This means that one obtains the coefficients a_{rs} in the addition theorem of an arbitrary genus φ by $a_{rs} = \varphi(\underline{a}_{rs})$. In this sense, the above formula yields a universal addition theorem (cf. [Qu69]).

The derivative of the logarithm $\underline{g}$ belonging to $\underline{\varphi}$ is, by our discussion in section 1.6,

$$\underline{g}'(\varphi) = \sum_{n=0}^{\infty} P_n(\mathbb{C}) \cdot y^n.$$

Milnor (cf. [Mi65]) has introduced some interesting manifolds, with whose help the $\underline{a}_{rs}$ can be determined.

In the following $\mathbb{P}_i$, resp. $\mathbb{P}_j$, shall always denote the complex projective space of (complex) dimension i, resp. j. We consider $\mathbb{P}_i \times \mathbb{P}_j$: The cohomology of the components is generated as rings by $u \in H^2(\mathbb{P}_i; \mathbb{Z})$ with $u^{i+1} = 0$ and $v \in H^2(\mathbb{P}_j; \mathbb{Z})$ with

$v^{j+1} = 0$. We further identify u, resp. v, with the pulled-back classes under the projection onto the components, therefore $u, v \in H^2(\mathbb{P}_i \times \mathbb{P}_j; \mathbb{Z})$ and $H^*(\mathbb{P}_i \times \mathbb{P}_j; \mathbb{Z}) = \mathbb{Z}[u,v]/(u^{i+1}, v^{j+1})$. Let $x = (x_0 : \ldots : x_i)$ and $y = (y_0 : \ldots : y_j)$ be the homogeneous coordinates on $\mathbb{P}_i$, resp. $\mathbb{P}_j$. A polynomial $P(x,y)$ has double degree (a,b) if it is homogeneous in x of degree a, and homogeneous in y of degree b, so that $P(\lambda \cdot x, \mu \cdot y) = \lambda^a \mu^b P(x,y)$. Then the equation $P(x,y) = 0$ defines a hypersurface in $\mathbb{P}_i \times \mathbb{P}_j$. This is smooth exactly if the partial derivatives of P with respect to the x_α and y_β vanish only for $x = y = 0$. Such a smooth hypersurface of double degree (a,b) represents a cohomology class which depends only on a and b, namely $a \cdot u + b \cdot v$. We consider the smooth hypersurface $H_{ij} \subset \mathbb{P}_i \times \mathbb{P}_j$ of double degree $(1,1)$, defined by $x_0 y_0 + \cdots + x_k y_k = 0$ ($k = \min\{i,j\}$). This has real dimension $2(i+j-1)$ and represents the cohomology class $u + v$. We therefore obtain

$$\varphi(H_{ij}) = H_{ij} = \sum_{r,s \geq 0} \underline{a}_{rs} \big(\underbrace{u, \ldots, u}_{r}, \underbrace{v, \ldots, v}_{s} \big)$$

$$= \sum_{r=0}^{i} \sum_{s=0}^{j} \underline{a}_{rs} \cdot \mathbb{P}_{i-r} \cdot \mathbb{P}_{j-s},$$

since the virtual submanifolds $\big(\underbrace{u, \ldots, u}_{r} \big)$ resp. $\big(\underbrace{v, \ldots, v}_{s} \big)$ in $\mathbb{P}_i$, resp. $\mathbb{P}_j$ are represented by $\mathbb{P}_{i-r} \subset \mathbb{P}_i$, resp. $\mathbb{P}_{j-s} \subset \mathbb{P}_j$.

With the notations $\underline{F}(y_1, y_2) = \sum_{r,s \geq 0} \underline{a}_{rs} \cdot y_1^r y_2^s$ and $H(y_1, y_2) = \sum_{i,j \geq 0} H_{ij} \cdot y_1^i y_2^j$ we have the following

Proposition: $H(y_1, y_2) = \underline{F}(y_1, y_2) \cdot \underline{g}'(y_1) \cdot \underline{g}'(y_2)$.

Proof:

$$H(y_1, y_2) = \sum_{i,j \geq 0} H_{ij} \cdot y_1^i y_2^j$$

$$= \sum_{i,j \geq 0} \Big(\sum_{r=0}^{i} \sum_{s=0}^{j} \underline{a}_{rs} \cdot \mathbb{P}_{i-r} \cdot \mathbb{P}_{j-s} \cdot y_1^i y_2^j \Big)$$

$$= \sum_{r,s \geq 0} \Big(\underline{a}_{rs} \cdot y_1^r y_2^s \cdot \sum_{i \geq r} \mathbb{P}_{i-r} \cdot y_1^{i-r} \cdot \sum_{j \geq s} \mathbb{P}_{j-s} \cdot y_2^{j-s} \Big)$$

$$= \underline{F}(y_1, y_2) \cdot \underline{g}'(y_1) \cdot \underline{g}'(y_2). \qquad \qquad \square$$

Remark: This is an identity in the cobordism ring. It holds also over $\mathbb{Z}$. Since the power series $\underline{g}'(y)$ is invertible over $\mathbb{Z}$, one can express the $\underline{a}_{rs}$ in terms of the H_{ij} and the projective spaces. Thus one obtains concrete formulas, e.g.:

$$
\begin{aligned}
H_{0,1} &= \underline{a}_{0,1} + \mathbb{P}_1, & \underline{a}_{0,1} &= H_{0,1} - \mathbb{P}_1, \\
H_{1,1} &= \underline{a}_{1,1} + 2\underline{a}_{0,1} \cdot \mathbb{P}_1 + \mathbb{P}_1^2, & \underline{a}_{1,1} &= H_{1,1} - 2 \cdot H_{0,1} \cdot \mathbb{P}_1 + \mathbb{P}_1^2.
\end{aligned}
$$

Now let φ be an elliptic genus, so that with $y = f(x)$ we have the formulas

$$f'^2 = P(y) = 1 - 2\delta \cdot y^2 + \varepsilon \cdot y^4, \quad g'(y) = \frac{1}{\sqrt{P(y)}},$$

$$F(y_1, y_2) = \frac{y_1 \cdot \sqrt{P(y_2)} + y_2 \cdot \sqrt{P(y_1)}}{1 - \varepsilon \cdot y_1^2 y_2^2}.$$

From this we obtain the following

Corollary:

$$\sum_{i,j \geq 0} \varphi(H_{ij}) \cdot y_1^i y_2^j = \left(\frac{y_1}{\sqrt{P(y_1)}} + \frac{y_2}{\sqrt{P(y_2)}} \right) \cdot \left(1 + \varepsilon y_1^2 y_2^2 + \varepsilon^2 y_1^4 y_2^4 + \cdots \right). \qquad \square$$

Corollary: *Let $j \geq i$ without restriction. Then*

$$\varphi(H_{ij}) = \begin{cases} 0, & i \equiv 1 \ (2), \\ \varepsilon^{i/2} \cdot \varphi(P_{j-i-1}(\mathbb{C})), & i \equiv 0 \ (2). \end{cases}$$

Proof: One can obtain the summand $\varphi(H_{ij}) \cdot y_1^i y_2^j$ in the above equation for $j \geq i$ only by multiplying $y_2 / \sqrt{P(y_2)} = y_2 \cdot g'(y_2)$ with $\left(1 + \varepsilon y_1^2 y_2^2 + \varepsilon^2 y_1^4 y_2^4 + \cdots \right)$. The only term that yields the monomial $y_1^i y_2^j$ for even i is $a_{j-i} \cdot y_2^{j-i} \cdot \varepsilon^{i/2} y_1^i y_2^i$ where a_k is the k-th coefficient of the power series

$$\frac{y}{\sqrt{P(y)}} = y \cdot g'(y) = \sum_{k=1}^{\infty} \varphi(\mathbb{P}_{k-1}) y^k.$$

For odd i there is no monomial $y_1^i y_2^j$. $\qquad \square$

Remark: Because $H_{ij} = H_{ji}$, all cases are included. In particular, $\varphi(H_{ij})$ has the following values for special i and j:

1) j even, $j \geq i$: value 0,
2) $|i - j|$ even: value 0,
3) i even, $j = i + 1$: value $\varepsilon^{i/2}$.

Corollary: $\hat{A}(H_{ij}) = 0$ *for* $i, j > 0$.

Proof: The $\hat{A}$-genus is the elliptic genus for $\varepsilon = 0$ (with the normalization $\delta = -\frac{1}{8}$). So the vanishing on H_{ij} for $i, j > 0$ follows from the above corollary. $\qquad \square$

4 Multiplicativity in fibre bundles

4.1 The signature and the L-genus

The Milnor manifolds $H_{ij} \subset \mathbb{P}_i \times \mathbb{P}_j$ of the last section are total spaces of a beautiful fibre bundle. To see this, we consider the projection of $\mathbb{P}_i \times \mathbb{P}_j$ onto $\mathbb{P}_i$. This induces for $i \leq j$ a fibration of H_{ij} over $\mathbb{P}_i$ with fibre $\mathbb{P}_{j-1}$, as one sees directly from the equation for H_{ij}. The manifold H_{ij} is therefore the total space of a projective bundle over $\mathbb{P}_i$. Since for even j every genus $\varphi(\mathbb{P}_{j-1})$ is zero, and since in the last section we saw that for elliptic genera also $\varphi(H_{ij}) = 0$ for even j with $j \geq i$, it follows that elliptic genera behave multiplicatively for these fibre bundles:

$$\varphi(H_{ij}) = \varphi(\mathbb{P}_i) \cdot \varphi(\mathbb{P}_{j-1}) \qquad \text{for } j \text{ even, } j \geq i.$$

Remark: This multiplicativity does not hold for all i, j. For $j = i + 1$ and i even one obtains $\varphi(H_{ij}) = \varepsilon^{i/2}$ (cf. section 3.2), which can very well differ from $\varphi(\mathbb{P}_i)^2$. E.g., for $i = 2$ the condition for multiplicativity is

$$\varepsilon = \varphi(H_{2,3}) = \varphi(\mathbb{P}_2)^2 = \delta^2,$$

which characterizes up to normalization the L-genus.

We shall now consider more closely the L-genus in connection with multiplicativity. The L-genus is the elliptic genus with $\varepsilon = \delta^2 = 1$. It satisfies $Q(x) = x/\tanh(x)$ and $f' = 1 - f^2$. Since $g'(y) = (1 - y^2)^{-1}$, the L-genus is characterized by $L(\mathbb{P}_{2n}) = 1$. Without normalization, it would give for $\mathbb{P}_{2n}$ the value δ^n.

For a $4n$-dimensional, oriented, compact manifold M there is in a natural way a quadratic form on $H^{2n}(M; \mathbb{R})$, given by $a \mapsto (a \cdot a)[M]$. This can be diagonalized with respect to a suitable basis of $H^{2n}(M; \mathbb{R})$, say with p diagonal entries 1 and q diagonal entries -1. Then p and q are uniquely determined. Due to Poincaré duality, this quadratic form is non-degenerate, hence $p + q = \dim H^{2n}(M; \mathbb{R}) = b_{2n}$, the middle Betti number of the manifold. One defines the signature of M as the signature of the quadratic form, i.e. $\operatorname{sign}(M) := p - q$.

The L-genus yields precisely this important invariant of a manifold M^{4n}:

Signature Theorem (Hirzebruch):

$$L(M^{4n}) = \operatorname{sign}(M^{4n}).$$

Proof (cf. [Hi56]): The signature is a cobordism invariant, which is compatible with the ring structure (disjoint union and Cartesian product) in the cobordism ring $\Omega \otimes \mathbb{Q}$. Moreover, $\mathrm{sign}\,(\mathbb{P}_{2n}) = 1$ and thus, since the $\mathbb{P}_{2n}$ from a basis sequence, $\mathrm{sign} = L$.

Now in 1957 Chern, Hirzebruch and Serre proved the following result (cf. [ChHiSe57]):

Theorem: *Let E be the total space of a differentiable fibre bundle with fibre F and base space B. Suppose that E, B, F are connected, compact, oriented manifolds, and that the fundamental group $\pi_1(B)$ acts trivially on $H^*(F; \mathbb{R})$. Then the signature is multiplicative in this fibre bundle, i.e.* $\mathrm{sign}\,(E) = \mathrm{sign}\,(B) \cdot \mathrm{sign}\,(F)$.

Remark: For simply connected base spaces B the theorem is naturally always applicable, since then $\pi_1(B) = 0$. Without the assumption on $\pi_1(B)$ the theorem is false (cf. section 4.6)

Question: Which genera are multiplicative under the above assumptions? (The Euler number is multiplicative in arbitrary fibre bundles, but is not an oriented cobordism invariant.)

Theorem (Borel, Hirzebruch): *The signature is the only genus which is multiplicative in fibre bundles* (E, B, F) *with simply connected B and takes the value* 1 *on* $P_2(\mathbb{C})$.

Proof: As stated above, for $i \leq j$ the H_{ij} are total spaces of fibre bundles over $\mathbb{P}_i$ with fibre $\mathbb{P}_{j-1}$. For $i = 2$ and $j \neq 1$ odd, we have a sequence of 8, 12, 16, ...-dimensional manifolds, which are all fibered over $\mathbb{P}_2$. We show later in this section that $\mathbb{P}_2$ and the $H_{2,2k+1}$ ($k = 1, 2, \dots$) form a basis sequence of the cobordism ring. Now let φ be multiplicative as above. Since $\varphi(\mathbb{P}_2) = L(\mathbb{P}_2)$, the genus φ coincides with L on all manifolds of dimension 4 (these are generated in $\Omega \otimes \mathbb{Q}$ by $\mathbb{P}_2$). Then also $\varphi(H_{2,3}) = L(H_{2,3})$, since $H_{2,3}$ is fibered over $\mathbb{P}_2$ with fibre $\mathbb{P}_2$ and φ and L are both multiplicative ($\mathbb{P}_2$ is simply connected). Now φ and L coincide on all manifolds up to dimension 8 (they are generated by $\mathbb{P}_2$ and $H_{2,3}$). Since all $H_{2,2k+1}$ fibre analogously into lower-dimensional projective spaces with base $\mathbb{P}_2$ (simply connected), one shows easily by induction that φ and L coincide on the entire cobordism ring. For the proof see also [BoHi59].

In order to show that the $H_{2,2k+1}$ together with $\mathbb{P}_2$ form a basis sequence, we consider the number $s\big(M^{4k}\big)$ which is defined as follows:

$$s\big(M^{4k}\big) := \big(x_1^{2k} + \cdots + x_k^{2k}\big)\big[M^{4k}\big],$$

if $p(M) = \big(1 + x_1^2\big) \cdots \big(1 + x_k^2\big)$. The number s is, on manifolds of fixed dimension $4k$, the genus for $Q(x) = 1 + x^{2k}$. We shall also denote this genus by s_k.

Lemma: $s(\mathbb{P}_{2k}) = 2k + 1$.

Proof: We have

$$p(\mathbb{P}_{2k}) = \big(1 + g^2\big)^{2k+1}$$

$$\Rightarrow \quad s(\mathbb{P}_{2k}) = s_k(\mathbb{P}_{2k}) = \big(1 + g^{2k}\big)^{2k+1}[\mathbb{P}_{2k}].$$

The coefficient of g^{2k} in this expression is $2k+1$ and

$$(2k+1) \cdot g^{2k}[\mathbb{P}_{2k}] = 2k+1.$$

Remark: For a product $M^{4k} = U \times V$ with $\dim U$, $\dim V < 4k$ we have $s(M) = 0$, since:

$$s(M) = s_k(M) = s_k(U) \cdot s_k(V) = 0.$$

Theorem (Thom): *A sequence M^4, $M^8, \ldots$ of manifolds is a basis sequence (i.e. $\Omega \otimes \mathbb{Q} = \mathbb{Q}[M^4, M^8, \ldots]$) if and only if $s(M^{4k})$ is different from zero for all k (cf. [Th54]).*

Proof: In $\Omega \otimes \mathbb{Q}$ each manifold M^{4k} is a polynomial in the projective spaces $\mathbb{P}_{2j}$, which are known to be a basis sequence. Evidently $\mathbb{P}_{2k}$ enters only linearly (with a coefficient a_k) in this polynomial. The M^{4k} therefore form a basis sequence if and only if $a_k \neq 0$ for all k (proof by induction). However, by the above remark,

$$s(M^{4k}) = s_k(M^{4k}) = a_k \cdot s_k(\mathbb{P}_{2k}) = a_k \cdot (2k+1).$$

Proposition: *If $i, j \geq 2$ with $\dim_\mathbb{C} H_{ij} = i + j - 1$ even, then*

$$s(H_{ij}) = -\binom{i+j}{i} \neq 0.$$

Proof: By construction of the $H_{ij} \subset \mathbb{P}_i \times \mathbb{P}_j$, the results on virtual submanifolds yield:

$$p(H_{ij}) = i^*\left((1+u^2)^{i+1} \cdot (1+v^2)^{j+1} \cdot (1+(u+v)^2)^{-1}\right)$$

$$\Rightarrow \quad s(H_{ij}) = s_{(i+j-1)/2}(H_{ij})$$

$$= i^*\left((1+u^{i+j-1})^{i+1} \cdot (1+v^{i+j-1})^{j+1} \cdot (1+(u+v)^{i+j-1})^{-1}\right)[H_{ij}].$$

Since for $j \geq 2$ (hence $i + j - 1 \geq i + 1$) it follows that $u^{i+j-1} = 0$ (analogously $i \geq 2$ implies $v^{i+j-1} = 0$), there only remains:

$$s(H_{ij}) = i^*\left((1+(u+v)^{i+j-1})^{-1}\right)[H_{ij}]$$

$$= i^*\left(1-(u+v)^{i+j-1}\right)[H_{ij}]$$

$$= \left((1-(u+v)^{i+j-1}) \cdot (u+v)\right)[\mathbb{P}_i \times \mathbb{P}_j]$$

$$= -\binom{i+j}{i}(u^i v^j)[\mathbb{P}_i \times \mathbb{P}_j]$$

$$= -\binom{i+j}{i}.$$

Here we have twice dropped terms outside of the correct cohomology dimension in the calculation.

Corollary:

1) *The manifolds $\mathbb{P}_2$ and $H_{2,2k+1}$ for $k \geq 1$ form a basis sequence.*
2) *The manifolds $H_{3,2k-2}$ for $k \geq 1$ form a basis sequence.*

Proof: Since both sequences contain a manifold of dimension $4k$ for each k, it suffices to show that s does not vanish on these manifolds. This is clear for $\mathbb{P}_2 \cong H_{3,0}$, and follows for $H_{2,2k+1}$ and $H_{3,2k-2}$ ($k > 1$) from the theorem.

One can even construct a basis sequence of the cobordism ring over $\mathbb{Z}$ (modulo torsion) from the complex projective spaces and the H_{ij}. This follows from (cf. [St68], p. 207)

Theorem (Milnor): *A sequence of manifolds M^{4k} is a basis sequence over $\mathbb{Z}$ of Ω modulo torsion if it satisfies:*

$$s\left(M^{4k}\right) = \begin{cases} \pm p, & 2k + 1 = p^r, \ p \text{ prime}, \\ \pm 1, & \text{otherwise}. \end{cases}$$

Corollary: *A number $n \in \mathbb{Z}$ can be written as $s\left(M^{4k}\right)$ if and only if $2k + 1$ is not a prime power, or $n \equiv 0 \ (p)$ for $2k + 1 = p^r$.*

4.2 Algebraic preliminaries

Up to now, for a complex bundle E over a manifold X we have always factored the Chern classes $c_i(E)$ formally, i.e. $c(E) = \prod_{i=1}^{n} (1 + x_i)$. In the next section we will carry out the factorization geometrically (the splitting principle, cf. [Hi56] or [BoTu82]), and thereby locate these x_i in the cohomology ring of a suitable manifold. We first give a few considerations of an algebraic nature.

We consider the polynomial ring $S = \mathbb{Z}[c_1,\ldots,c_n]$ in the indeterminates c_i of weight $2i$. Later we shall substitute for the c_i the Chern classes of a bundle, and so obtain a mapping of S into the cohomology ring of the manifold X, where naturally relations can appear. Further, let $R = \mathbb{Z}[x_1,\ldots,x_n]$ be the polynomial ring in the indeterminates x_i of weight 2. By means of $c_i \mapsto \sigma_i(x_1,\ldots,x_n)$ we obtain a degree preserving injection of S into R (σ_i is the i-th elementary symmetric function). We identify the polynomial ring S with its image in R. Hence S becomes a subring of R, so that R is a graded S-module. We want to determine a basis of this module.

We have:

$$(t + x_1) \cdots (t + x_n) = t^n + t^{n-1} \cdot c_1 + \cdots + c_n,$$

and for $t = -x_1$ we get

$$0 = x_1^n - x_1^{n-1}c_1 \pm \cdots + (-1)^n c_n.$$

Hence $\{1, x_1, \ldots, x_1^{n-1}\}$ is a basis of the S-module $S[x_1] \subset R$. The product $(t + x_2) \cdots (t + x_n)$ is a polynomial in t, whose coefficients are polynomials in the c_j and x_1, for:

$$(t + x_2) \cdots (t + x_n) = t^{n-1} + t^{n-2} \cdot \widetilde{c}_1 + \cdots + \widetilde{c}_{n-1},$$

where $\widetilde{c}_i$ can be recursively determined by

$$c_k(x_1, \ldots, x_n) = \widetilde{c}_k(x_2, \ldots, x_n) + x_1 \widetilde{c}_{k-1}(x_2, \ldots, x_n) \quad \text{and} \quad \widetilde{c}_0 := 1.$$

We therefore obtain inductively as basis elements of R over S all $n!$ monomials of the form $x_1^{i_1} \cdots x_n^{i_n}$ with $0 \leq i_j \leq n - j$ (in particular, always $i_n = 0$, since $x_n = c_1 - x_1 - \cdots - x_{n-1}$). Thus there is precisely one basis element of highest weight, namely $x_1^{n-1} \cdot x_2^{n-2} \cdots x_{n-1}$ with weight $n(n-1)$. Let $P \in R$; then P can be written uniquely as an S-linear combination of the above basis elements. We are especially interested in the coefficient $\widetilde{\rho}(P) \in S$ of the unique basis element of highest weight. We shall now determine this for the polynomial $P = \prod_{i>j}(x_i - x_j)$. Because P and all the basis elements are homogeneous, this also holds for the coefficients $s_{i_1,\ldots,i_n}$ in the basis representation. Since P has weight $n(n-1)$, the coefficient $s_{i_1,\ldots,i_n}$ has weight $n(n-1) - 2\sum_{j=1}^{n} i_j$. Therefore the coefficient $\widetilde{\rho}(P)$ has weight zero, hence $\widetilde{\rho}(P) \in \mathbb{Z}$, while all other coefficients have strictly positive weight. Therefore we can determine $\widetilde{\rho}(P)$ by computing P modulo the ideal $I = (c_1, \ldots, c_n)$.

To do this, we consider the polynomial

$$(t + x_1) \cdots (t + x_n) = t^n + t^{n-1} \cdot c_1 + \cdots + c_n;$$

putting $t = 1 - x_1$ gives:

$$(1 + x_2 - x_1) \cdots (1 + x_n - x_1) = (1 - x_1)^n + (1 - x_1)^{n-1} \cdot c_1 + \cdots + c_n.$$

Since the ideal I is homogeneous, we can consider the terms of equal weight modulo I and obtain as components of highest weight of these expressions:

$$(x_2 - x_1) \cdots (x_n - x_1) \equiv n \cdot (-x_1)^{n-1} \quad (I).$$

By induction we obtain

$$\prod_{i>j}(x_i - x_j) \equiv n! \cdot (-x_1)^{n-1}(-x_2)^{n-2} \cdots (-x_{n-1}) \quad (I)$$

$$\equiv n! \cdot (-1)^{n(n-1)/2} x_1^{n-1} \cdots x_{n-1} \quad (I),$$

which gives

$$\widetilde{\rho}\Big(\prod_{i>j}(x_i - x_j)\Big) = (-1)^{n(n-1)/2} \cdot n!.$$

For an arbitrary polynomial P in R we now put

$$\rho(P) := \widetilde{\rho}(P) \cdot (-1)^{n(n-1)/2}.$$

With the notation $\mathrm{sgn}\,(s)$ for the sign of a permutation s, we have the

Proposition: $\rho(P) = \left(\sum_{s \in S_n} \mathrm{sgn}\,(s) \cdot s(P)\right) / \prod_{i>j} (x_i - x_j)$.

Proof: Let Q be the numerator of the right side; then for each transposition $\tau_{ij} \in S_n$ we have $\tau_{ij}(Q) = -Q$. Polynomials with this property contain the factor $x_i - x_j$. Therefore the quotient on the right side is a polynomial. In particular, it vanishes if the weight of P is smaller than $n(n-1)$. Since the quotient is also invariant under the action of S_n, it is a polynomial in c_1 through c_n. Moreover, the right side is S-linear. Since this is also true of the left side, it suffices to verify the equation on an S-basis of R. For all our basis elements below weight $n(n-1)$ both sides give zero. In place of the basis element of highest weight we use the polynomial $P = \prod_{i>j} (x_i - x_j)$, in whose basis representation the summand $(-1)^{n(n-1)/2} n! \cdot x_1^{n-1} \cdots x_{n-1}$ enters, as has just been verified. Therefore $\rho(P) = n!$. On the other hand, $s(P) = \mathrm{sgn}\,(s) \cdot P$, hence all summands are equal to the denominator, and therefore on the right side we also obtain $n!$. $\qquad\qquad$ ▢

4.3 Topological preliminaries

Let X and Y be compact, oriented manifolds. In the following we always consider, if nothing different is said, (co-) homology groups with coefficients in $\mathbb{Z}$. By Poincaré duality, $H^*(X) \cong H_*(X)$; more precisely, $H^i(X) \cong H_{n-i}(X)$, where $n = \dim X$. Let $f : X \to Y$ be a continuous map. Then f induces maps $f^* : H^*(Y) \to H^*(X)$ and $f_* : H_*(X) \to H_*(Y)$ on the cohomology, resp. homology level. Poincaré duality p now yields also a mapping $f_* : H^*(X) \to H^*(Y)$, more precisely

$$
\begin{array}{ccc}
H^i(X) & \xrightarrow{\;f_*\;} & H^{m-n+i}(Y) \\
\downarrow p & & \downarrow p \\
H_{n-i}(X) & \xrightarrow{\;f_*\;} & H_{n-i}(Y),
\end{array}
$$

where $m = \dim Y$. For $b \in H^n(X)$ we have $f_*(b) \in H^m(Y)$ and

$$b[X] = f_* b[Y],$$

where $[X]$ and $[Y]$ again denote the fundamental cycles of X and Y. Since $H^*(X)$ is a ring and $f^* : H^*(Y) \to H^*(X)$ is a ring homomorphism, $H^*(X)$ is also a module over $H^*(Y)$ and for $a \in H^*(Y), b \in H^*(X)$ we have

$$f_*(f^*(a) \cdot b) = a \cdot f_*(b),$$

i.e. f_* is an $H^*(Y)$-module homomorphism.

Let $f : X \to Y$ be a fibre bundle where the fibre A is a compact, oriented manifold. Then one has:

$$f_* : H^*(X) \to H^*(Y),$$
$$f_* : H^i(X) \to H^{i-\dim A}(Y).$$

Remark: In the differentiable case one can regard this mapping f_* in de Rham cohomology as integration of differential forms over the fibres.

From now on, let X and Y as well as $f : X \to Y$ be differentiable. We define a sub-bundle X^Δ of the tangent bundle TX of X. The fibre of X^Δ over a point $x \in X$ is formed from all tangent vectors at the point x which are tangent to the fibre A through x. We call this bundle X^Δ the bundle along the fibres. The restriction of X^Δ to a fibre is isomorphic to the tangent bundle to A. We have:

$$TX \cong X^\Delta \oplus f^*TY.$$

Example: For a product $X \cong Y \times A$, $TX \cong \pi^*TA \oplus f^*TY$, where π is the projection of X onto A.

Now let $Q(x)$ be an even power series in x with coefficients some integral domain R over $\mathbb{Q}$. The corresponding multiplicative sequence for a bundle E is then $K(E) = K_Q(E) = \sum_{i=0}^\infty K_i(p_1, \ldots, p_i)$, where $p_i = p_i(E)$. Then we have:

$$K(TX) = K(X^\Delta) \cdot f^*K(TY)$$
$$\Rightarrow \quad K(X) = K(TX)[X]$$
$$= f_*\big(f^*K(TY) \cdot K(X^\Delta)\big)[Y]$$
$$= \big(K(TY) \cdot f_*K(X^\Delta)\big)[Y].$$

By evaluation on a point in Y one obtains:

$$f_*K(X^\Delta) = K(A) \cdot 1 + (\text{higher terms})$$
$$\Rightarrow \quad K(X) = \big(K(TY) \cdot (K(A) \cdot 1 + (\text{higher terms}))\big)[Y].$$

The higher terms must be present if the genus K does not behave multiplicatively for the fibre bundle. Therefore we make the following

Definition: *The genus corresponding to the power series Q is called **strictly multiplicative** in the fibre bundle $f : X \to Y$ with fibre A if $f_*K(X^\Delta) = K(A) \cdot 1 \in H^0(Y; R)$.*

Thus we obtain the

Proposition: *If K is strictly multiplicative, then $K(X) = K(Y) \cdot K(A)$.*

4.4 The splitting principle

Let $E \to Y$ be a complex vector bundle over a compact, oriented, differentiable manifold Y. The flag manifold $F(V)$ belonging to a complex vector space V of dimension n consists of all flags of subspaces of V of length n. For the vector space $V = \mathbb{C}^n$ with its standard basis we have $F(n) := F(\mathbb{C}^n) = \mathrm{U}(n)/T^n$, where $T^n = (\mathrm{U}(1))^n \cong (S^1)^n$ is the standard maximal torus in $\mathrm{U}(n)$. Notice further that two flag manifolds over different vector spaces of equal dimension, but lacking canonical isomorphism, are not canonically isomorphic.

We now consider the bundle $\sigma_n : F_n \to Y$, which has as fibre A over each point $x \in Y$ the flag manifold of the corresponding fibre of E (F_n is the $F(n)$-bundle associated to E, having structure group $\mathrm{U}(n)$). In addition, we form in the same manner still further $\mathrm{U}(n)$-bundles $\sigma_i : F_i \to Y$, having the fibres

$$\mathrm{U}(n)/\big(T^i \times \mathrm{U}(n-i)\big) = \mathrm{U}(n)/\big(\mathrm{U}(1)^i \times \mathrm{U}(n-i)\big)$$

(increasing flags of length i). Between consecutive bundles there is a natural projection $\tau_i : F_i \to F_{i-1}$ with fibre

$$\frac{\mathrm{U}(n)/\big(T^i \times \mathrm{U}(n-i)\big)}{\mathrm{U}(n)/\big(T^{i-1} \times \mathrm{U}(n-i+1)\big)} \cong \frac{\mathrm{U}(n-i+1)}{T^1 \times \mathrm{U}(n-i)} \cong P_{n-i}(\mathbb{C}).$$

With this definition we have $\sigma_i = \tau_1 \tau_2 \cdots \tau_i$, and we define the projection π_i by $\pi_i := \tau_{i+1} \cdots \tau_{n-1}\tau_n$, so that $\pi_i : F_n \to F_i$, with respect to which F_n is a bundle with fibre $\mathrm{U}(n-i)/T^{n-i} \cong F(n-i)$.

A point in a fibre of F_1 canonically represents a line in the corresponding fibre of E. Therefore F_1 is the projectivized bundle $\mathbb{P}E$. If one pulls back E to F_1 by means of σ_1, then over each point of F_1 one has a distinguished line in the fibre of the pulled-back bundle over this point, and can so split off a line bundle L_1,

$$\sigma_1^*(E) = L_1 \oplus E_1.$$

Now a point in a fibre of F_2 that lies over some point L in F_1, canonically represents a flag of length 2 which consists of the line given by the point L in F_1 and a plane containing this line. So this flag can alternatively be given by a line in the complement of L, i.e. a line in the fibre of E_1 over L in F_1. Therefore F_2 is the projectivized bundle $\mathbb{P}E_1$ and we can proceed inductively, and $\sigma_i^* E$ automatically splits off i line bundles,

$$\sigma_i^* E = L_1 \oplus \cdots \oplus L_i \oplus E_i,$$

where E_i is a vector bundle of rank $n - i$.

Thus we have the following diagram:

$$
\begin{array}{ccc}
L_1 \oplus \cdots \oplus L_n & \longrightarrow & F_n \\
\downarrow & & \downarrow \tau_i \\
L_1 \oplus \cdots \oplus L_i \oplus E_i & \longrightarrow & F_i \\
\downarrow & & \downarrow \\
L_1 \oplus E_1 & \longrightarrow & F_1 \\
\downarrow & & \downarrow \sigma_1 \\
E & \xrightarrow{\ f\ } & Y.
\end{array}
$$

The line bundles L_i restricted to a fibre of $F_i \to F_{i-1}$ are tautological bundles. Let them have first Chern classes $x_i \in H^2(F_i; \mathbb{Z})$. One deduces (consider the Leray spectral sequence, or more simply apply the Leray-Hirsch theorem) that $H^*(F_i; \mathbb{Z})$ is freely generated as a module over $H^*(F_{i-1}; \mathbb{Z})$ by $1, x_i, \ldots, x_i^{n-i}$, and the mapping $\tau_i^* : H^*(F_{i-1}; \mathbb{Z}) \to H^*(F_i; \mathbb{Z})$ is injective; we can therefore consider the cohomology rings as subrings of $H^*(F_n; \mathbb{Z})$, and with this identification we have:

$$
\sigma_n^* E = L_1 \oplus \cdots \oplus L_n,
$$
$$
c(E) = \sigma_n^* c(E) = (1 + x_1) \cdots (1 + x_n).
$$

Let F_i^Δ be the bundle along the fibres of the fibre bundle $\tau_i : F_i \to F_{i-1}$. Then we have the following (cf. [Hi56], Theorem 13.1.1 e)

Lemma: $F_i^\Delta \cong E_i \otimes L_i^{-1}$.

Proof: Restricted to the fibres (i.e. in the special case that Y is a point) we have $F_1^\Delta = T\mathbb{P}_{n-1}$. One has the exact sequence of vector bundles

$$
0 \to 1 \to n \cdot H^* \to T\mathbb{P}_{n-1} \to 0.
$$

Here, 1 is a trivial line bundle and H is the tautological line bundle, hence is the restriction of L_1 to the fibre. We have

$$
n \cdot H^* = n \cdot 1 \otimes H^* = \left(H \oplus \widetilde{E}_1\right) \otimes H^*,
$$

where $H \otimes H^*$ is the image of the trivial line bundle in the above sequence and $\widetilde{E}_1$ is the restriction of E to the fibre. Since the sequence splits, it follows that

$$
T\mathbb{P}_{n-1} \cong \widetilde{E}_1 \otimes H^*.
$$

From the functoriality of the construction, the statement for F_1^Δ now follows. For F_i^Δ with $i > 1$ one obtains the result inductively, by carrying out the construction for the bundle $E_{i-1} \to F_{i-1}$.

From the lemma we therefore obtain the following formulas:

$$\pi_i^* F_i^\Delta = L_{i+1} \otimes L_i^{-1} \oplus \cdots \oplus L_n \otimes L_i^{-1},$$

$$\pi_i^* c(F_i^\Delta) = (1 + x_{i+1} - x_i) \cdots (1 + x_n - x_i).$$

For the bundle F^Δ along the fibres of the map σ_n we therefore have

$$c(F^\Delta) = c(\pi_1^* F_1^\Delta \oplus \cdots \oplus \pi_n^* F_n^\Delta) = \prod_{i>j} (1 + x_i - x_j).$$

In the algebraic preliminaries we obtained from

$$(t + x_1) \cdots (t + x_n) = t^n + t^{n-1} \cdot c_1 + \cdots + c_n,$$

by the substitution $t = 1 - x_1$, the identity

$$(1 + x_2 - x_1) \cdots (1 + x_n - x_1) = (1 - x_1)^n + (1 - x_1)^{n-1} \cdot c_1 + \cdots + c_n.$$

We apply this to the total Chern class of F_1^Δ and see that

$$c(F_1^\Delta) = (1 - x_1)^n + (1 - x_1)^{n-1} c_1 + \cdots + c_n \in H^*(F_1; \mathbb{Z}),$$

where c_i are the pulled-back Chern classes of E.

For the cohomology of the flag manifold over a vector space V of dimension n we have the following

Lemma: *The Euler number of the flag manifold is* $n!$.

Proof: We consider the trivial bundle V over a point. F_n is then the flag manifold of V and

$$H^*(F(V); \mathbb{Z}) = \mathbb{Z}[x_1, \ldots, x_n]/(c_1, \ldots, c_n).$$

Therefore, as a $\mathbb{Z}$-module $H^*(F(V); \mathbb{Z})$ has the basis elements $x_1^{i_1} \cdots x_n^{i_n}$ with $0 \le i_j \le n - j$. Since these all lie in even dimensions and their number is $n!$, the assertion follows, for the Euler number is equal to the alternating sum of the Betti numbers.

Alternatively, one can also use the multiplicativity of the Euler number. Since $e(\mathbb{P}_k) = k + 1$ and $F(n)$ is fibered into $\mathbb{P}_1, \ldots, \mathbb{P}_{n-1}$, we have

$$e(F(n)) = e(\mathbb{P}_1) \cdots e(\mathbb{P}_{n-1}) = n! .$$

Lemma: $\sigma_{n*}\left(\prod_{i>j}(x_i - x_j)\right) = n!$.

Proof: Since the dimension of $\prod_{i>j}(x_i - x_j)$ is exactly $n(n-1)$ and σ_{n*} lowers the dimension by $n(n-1)$, it suffices to evaluate the expression over a point P:

$$\sigma_{n*}\left(\prod_{i>j}(x_i - x_j)\right)[P] = \left(\prod_{i>j}(x_i - x_j)\right)[F(n)]$$

$$= c_n(F^\Delta)[F(n)] = e(F(n)) = n! .$$

Here we have used the fact that in the fibre over a point P, the bundle F^Δ is isomorphic to $TF(n)$.

4.5 Integration over the fibre

Now we shall turn our attention to the integration over the fibre.

Theorem: *With $Q(x) = x/f(x)$ and $f(x) = x + \cdots$ an odd power series, there holds for F_1^Δ:*

$$\sigma_{1*} K_Q\left(F_1^\Delta\right) = \sum_{i=1}^{n} \prod_{j \neq i} \frac{1}{f(x_j - x_i)}.$$

Proof: Let a be a symmetric expression in the x_i, therefore a polynomial in the c_j. With our identification $a = \sigma_n^*(a)$, for an element $b \in H^*(F_n; \mathbb{Z})$ we have

$$\sigma_{n*}(a \cdot b) = \sigma_{n*}(\sigma_n^* a \cdot b) = a \cdot \sigma_{n*} b.$$

In addition, σ_{n*} is naturally additive, hence is $\mathbb{Z}[c_1, \ldots, c_n]$-linear. Furthermore, the cohomology dimension of an element is lowered by $n(n-1)$ under the mapping σ_{n*}. Since $H^*(F_n; \mathbb{Z})$ is generated, as a module over $H^*(Y; \mathbb{Z})$, by the elements $x_1^{i_1} \cdots x_n^{i_n}$ with $0 \leq i_j \leq n - j$, and the mapping σ_{n*} is $\mathbb{Z}[c_1, \ldots, c_n]$-linear with $\sigma_{n*}\left(\prod_{i>j}(x_i - x_j)\right) = n!$, it coincides with the map ρ occurring in section 4.2.

If we apply the above to the bundle $E_1 \to F_1$, then we obtain (inductive construction of the F_i)

$$\pi_{1*}\left(\prod_{i>j \geqslant 2}(x_i - x_j)\right) = (n-1)!$$

$$\Rightarrow \quad (n-1)! \cdot \sigma_{1*}\left(K_Q\left(F_1^\Delta\right)\right)$$

$$= (n-1)! \cdot \sigma_{1*}\left(Q(x_2 - x_1) \cdots Q(x_n - x_1)\right)$$

$$= \sigma_{1*}\left(\pi_{1*}\left(\prod_{i>j \geqslant 2}(x_i - x_j)\right) \cdot Q(x_2 - x_1) \cdots Q(x_n - x_1)\right)$$

$$= \sigma_{n*}\left(\prod_{i>j \geqslant 2}(x_i - x_j) \cdot Q(x_2 - x_1) \cdots Q(x_n - x_1)\right)$$

$$= \sum_{s \in S_n} \frac{\operatorname{sgn}(s) \cdot s\left(\prod_{i>j \geqslant 2}(x_i - x_j) \cdot Q(x_2 - x_1) \cdots Q(x_n - x_1)\right)}{\prod_{i>j}(x_i - x_j)}$$

$$= \sum_{s \in S_n} \frac{s\left(\prod_{i>j \geqslant 2}(x_i - x_j) \cdot Q(x_2 - x_1) \cdots Q(x_n - x_1)\right)}{s\left(\prod_{i>j}(x_i - x_j)\right)}$$

$$= \sum_{s \in S_n} s\left(\frac{Q(x_2 - x_1) \cdots Q(x_n - x_1)}{(x_2 - x_1) \cdots (x_n - x_1)}\right)$$

$$= (n-1)! \sum_{i=1}^{n} \prod_{j \neq i} \frac{Q(x_j - x_i)}{(x_j - x_i)}$$

$$= (n-1)! \sum_{i=1}^{n} \prod_{j \neq i} \frac{1}{f(x_j - x_i)}.$$

Here we have used the fact that S_n is the disjoint union of the $S_{n-1} \cdot \tau_{1i}$, where τ_{1i} transposes 1 and i. As we know from the algebraic preliminaries, the result is in fact a polynomial in the c_j. ⌸

Corollary: *The genus belonging to the power series* $Q(x) = x/f(x)$ *is strictly multiplicative in bundles with fibre* $P_{n-1}(\mathbb{C})$ *if and only if the following identity holds:*

$$\sum_{i=1}^{n} \prod_{j \neq i} \frac{1}{f(x_j - x_i)} \equiv c.$$

This constant c *is then the value of the genus on* $P_{n-1}(\mathbb{C})$. ⌸

4.6 Multiplicativity and strict multiplicativity

We now want to find power series $f(x)$ which behave strictly multiplicatively in fibre bundles with fibre $P_{n-1}(\mathbb{C})$ (cf. [Oc87]). We have just seen that the genus associated to $Q(x) = x/f(x)$ is strictly multiplicative in such fibre bundles if and only if there holds:

$$K_Q(P_{n-1}(\mathbb{C})) \equiv \frac{1}{f(x_2 - x_1) \cdots f(x_n - x_1)} + \cdots + \frac{1}{f(x_1 - x_n) \cdots f(x_{n-1} - x_n)}.$$

Proposition: *For a genus* φ *the following are equivalent:*
1) φ *is an elliptic genus,*
2) φ *is strictly multiplicative in fibre bundles with fibre* $P_{2n-1}(\mathbb{C})$ *for all* n,
3) φ *is strictly multiplicative in fibre bundles with fibre* $P_3(\mathbb{C})$.

Proposition: *For a genus* φ *the following are equivalent:*
1) φ *is up to normalization (i.e. passage to* $f(\alpha x)/\alpha$ *) the* L-*genus,*
2) φ *is strictly multiplicative in fibre bundles with fibre* $P_n(\mathbb{C})$ *for all* n,
3) φ *is strictly multiplicative in fibre bundles with fibre* $P_2(\mathbb{C})$.

Remark: Notice 3) $\Rightarrow$ 2)!

Proof of the propositions: The proof of both propositions follows directly from the following lemmata. ⌸

Lemma: 1) $\Rightarrow$ 2) *for elliptic genera.*

Proof: For all $x_1, \ldots, x_n \in \mathbb{C}$, with x_i pairwise distinct modulo L, the function

$$h(x) := \frac{1}{f(x - x_1) \cdots f(x - x_n)}$$

is elliptic with respect to L for even n (cf. section 2.2). Therefore, according to the residue theorem we have:

$$0 = (-1)^{n-1} \sum_{x \in \mathbb{C}/L} \mathrm{res}_x\, h$$

$$= \frac{1}{f(x_2 - x_1) \cdots f(x_n - x_1)} + \cdots + \frac{1}{f(x_1 - x_n) \cdots f(x_{n-1} - x_n)}.$$

Since the x_i were arbitrary, this identity also holds for the formal power series. Further, since the coefficients of the power series f are polynomials in ε and δ, the above identity also holds for degenerate lattices on grounds of continuity.

For odd n, elliptic genera are in general not strictly multiplicative. For then the function considered above is only elliptic with respect to a lattice $\tilde{L}$ of index two in L. Therefore all poles appear doubled, but with opposite signs. Hence the whole expression on the right side cancels, and one obtains no information.

Lemma: 1) $\Rightarrow$ 2) *for the L-genus.*

Proof: For the function $\tanh(x)$, the fundamental mesh (period parallelogram) of the lattice L degenerates to a fundamental strip $H = \{z \in \mathbb{C} \mid 0 \leqslant \mathrm{Im}\,(z) \leqslant \pi i\}$ ($\tanh$ has the period πi). The pole set of $1/f(x)$ is $\mathbb{Z}\pi i$. For the symmetrized sum therefore holds:

$$\sum_{i=1}^{n} \prod_{j \neq i} \frac{1}{\tanh(x_j - x_i)} = \sum_{x \in H} \mathrm{res}_x \frac{(-1)^{n-1}}{\tanh(x - x_1) \cdots \tanh(x - x_n)}.$$

If one now attempts to calculate this residue sum by integrating the function

$$h(x) = \frac{1}{\tanh(x - x_1) \cdots \tanh(x - x_n)}$$

over the boundary of the fundamental strip H, one finds that:

$$h(x) \rightarrow \begin{cases} (-1)^n, & \text{for } x \rightarrow -\infty, \\ 1, & \text{for } x \rightarrow \infty. \end{cases}$$

Since the fundamental strip has width πi, integration over the fundamental strip gives for the sum of the residues the value 0 for even n and 1 otherwise, so is constant in every case.

Remark: 2) $\Rightarrow$ 3) is trivial.

Lemma: 3) $\Rightarrow$ 1).

Proof (exercise): Let $f(x)$ be an odd power series. Show by power series calculations:

i)   If f satisfies the identity

$$\sum_{i=1}^{3} \prod_{j \neq i} \frac{1}{f(x_j - x_i)} \equiv c,$$

then up to normalization $f(x) = \tanh(x)$.

ii)   If f satisfies the identity

$$\sum_{i=1}^{4}\prod_{j\neq i}\frac{1}{f(x_j - x_i)} \equiv 0,$$

then f belongs to an elliptic genus (namely show: f is uniquely determined by the coefficients of x^3 and x^5).

Remark: Alternatively, one can also prove this lemma by constructing a basis sequence in which all spaces are total spaces of fibre bundles, with fibre $P_2(\mathbb{C})$ for the L-genus, resp. $P_3(\mathbb{C})$ for the elliptic genera. One then argues inductively over the dimension of the manifold.

In order to deduce 1) from 2) without the characterization 3), one can proceed exactly as in the proof of the theorem of Borel and Hirzebruch in section 4.1. We recall the Milnor manifolds: These form for $i = 3$ and $j = 2k - 2, k \geqslant 1$ (thus $\dim H_{ij} = 4k$) a basis sequence (since then $s_k(H_{ij}) \neq 0$). In this situation, for $k \geqslant 3$ the manifold $H_{3,2k-2}$ fibres over $P_3(\mathbb{C})$ with fibre $P_{2k-3}(\mathbb{C})$ being a projective space of odd dimension. Since $H_{3,0} = P_2(\mathbb{C})$ and $H_{3,2} = H_{2,2}$, the implication 2) $\Rightarrow$ 1) is thereby proved.

In particular, we then have

$$\varphi(P_2(\mathbb{C})) = \delta, \quad \varphi(H_{2,3}) = \varepsilon, \quad \text{and} \quad \varphi(H_{3,2k-2}) = 0 \text{ for all } k \geqslant 3.$$

A more general result on multiplicativity in fibre bundles has been proved by Taubes and Bott-Taubes (cf. [Ta88], [BoTa89]). We give a reformulation using [Oc86], Proposition 1.

Theorem: *Let $M \to B$ be a fibre bundle with a compact, oriented, spin manifold F as fibre and compact, connected Lie group as structure group. Then every elliptic genus is multiplicative, i.e. $\varphi(M) = \varphi(F) \cdot \varphi(B)$.*

In order to make the difference between multiplicativity and strict multiplicativity vanish, we now have a

Theorem (Borel, Hirzebruch): *Let F be a fixed compact, oriented, differentiable manifold. Further, let ϑ_F be the category of all fibre bundles $\pi : M \to B$ with fibre F, where all manifolds considered are compact, oriented and differentiable and $\pi_1(B)$ operates trivially on $H^*(F;\mathbb{Q})$. If φ is a multiplicative genus, i.e. $\varphi(M) = \varphi(B) \cdot \varphi(F)$ for all these fibre bundles then φ is also strictly multiplicative in these fibre bundles, i.e. $\pi_*(K_\varphi(M^\Delta)) = \varphi(F) \cdot 1 \in H^0(B;\mathbb{Q})$.*

Proof (cf. [BoHi59]): According to Thom [Th54], the submanifolds of a compact, oriented, differentiable manifold generate the rational homology (a suitable multiple of every homology class can be represented by a $\mathbb{Z}$-linear combination of submanifolds).

We now proceed inductively as follows: Let φ be strictly multiplicative up to dimension k. Let K be the multiplicative sequence belonging to φ. For the bundle M^Δ along the fibres holds:

$$\pi_*\big(K\big(M^\Delta\big)\big) = \varphi(F) + a_k + a_{k+1} + \cdots,$$

where the $a_j \in H^j(B;\mathbb{Q})$. Now let X be a k-dimensional submanifold of B. The fundamental group $\pi_1(X)$ again acts trivially on $H^*(F;\mathbb{Q})$. We now restrict the fibre bundle to X. Then we have

$$\pi_*\big(K\big(M^\Delta|_X\big)\big) = \varphi(F) + a_k + 0$$

restricted to X. Now we have:

$$
\begin{aligned}
\varphi(X) \cdot \varphi(F) &= \varphi\big(\pi^{-1}(X)\big) \\
&= \big(K(\pi^*TX) \cdot K\big(M^\Delta|_X\big)\big)\big[\pi^{-1}X\big] \\
&= \big(K(TX) \cdot \pi_* K\big(M^\Delta|_X\big)\big)[X] \\
&= \big((1 + \cdots + \varphi(X) \cdot g_X) \cdot (\varphi(F) + a_k)\big)[X] \\
&= (\varphi(X) \cdot \varphi(F) \cdot g_X + a_k)[X] \\
&= \varphi(X) \cdot \varphi(F) + a_k[X] \qquad (\text{since } g_X[X] = 1) \\
\Rightarrow \quad a_k[X] &= 0 \quad \text{for all } X \\
\Rightarrow \quad a_k &= 0.
\end{aligned}
$$

Corollary: *The L-genus (signature) is strictly multiplicative in all differentiable bundles (with $\pi_1(B)$ acting trivially on $H^*(F;\mathbb{Q})$).*

Example: Let M and B be manifolds, $\dim M = 8$, $\dim B = 4$ and let M be total space of a fibre bundle with fibre F and base B (with $\pi_1(B)$ acting trivially on $H^*(F;\mathbb{Q})$). Then the vector bundle M^Δ has the Pontrjagin classes $p_1 \in H^4(M;\mathbb{Z})$ and $p_2 = e^2 \in H^8(M;\mathbb{Z})$. For the multiplicative sequence of the signature there holds in dimension 8:

$$L_2 = \big(7p_2 - p_1^2\big)/45.$$

Now $\pi_*\big(L_2\big(M^\Delta\big)\big) = 0$, for in view of the strict multiplicativity of the signature we have $\pi_*\big(L\big(M^\Delta\big)\big) = L(F) \cdot 1 \in H^0(B;\mathbb{Q})$, while $\pi_*\big(L_2\big(M^\Delta\big)\big)$ lives in $H^4(B;\mathbb{Q})$. Further, π_* is an isomorphism in the highest dimension, since Poincaré duality and $\pi_* : H_0(M;\mathbb{Z}) \to H_0(B;\mathbb{Z})$ are both isomorphisms. Hence over $\mathbb{Q}$ we have $L_2\big(M^\Delta\big) = 0$, so $7p_2 - p_1^2 = 0$ or

$$e^2 = p_1^2/7,$$

for the bundle along the fibres of an 8-dimensional manifold fibered over a 4-dimensional manifold.

Problem: Is there a bundle (with fibre dimension 4), for which the corresponding equation for the Pontrjagin classes does not hold? (In base dimension 4 there is no torsion, but in higher dimensions it is indeed possible.)

Example: Kodaira (cf. [Ko67]) has found a family of manifolds in the complex analytic category which shows that the trivial operation of $\pi_1(B)$ on F is a necessary hypothesis for our theorem. These manifolds are called Kodaira surfaces and are of general type. As complex surfaces, they have real dimension 4, and further they are total space of a fibre bundle with fibre and base a Riemann surface. If the L-genus were multiplicative in these fibre bundles, then on dimensional grounds the signature would be zero for these surfaces. However, Kodaira has calculated that the signature of these surfaces is strictly positive.

5 The Atiyah-Singer index theorem

5.1 Elliptic operators and elliptic complexes

Let X be a compact, differentiable manifold of dimension k, and let E and F be complex C^∞-vector bundles over X of rank m, resp. n. Denote by Γ the vector space of all C^∞-sections of a bundle, and let $D : \Gamma(E) \to \Gamma(F)$ be a $\mathbb{C}$-linear map. Each $f \in \Gamma(E)$ can be written over a trivializing neighborhood U of a point $x \in X$ as $f = (f_1, \ldots, f_m)$, where the f_i are C^∞-functions over U. In the same way the image $g = D(f)$ can be written locally as $g = (g_1, \ldots, g_n)$.

Definition: *D is called a **differential operator of order** p if there is an $m \times n$ matrix (d_{ij}) with*

$$(g_1, \ldots, g_n) = (f_1, \ldots, f_m) \cdot (d_{ij}),$$

where the d_{ij} are polynomials in the $\frac{\partial}{\partial x_r}$ of degree (at most) p with differentiable functions as coefficients.

Here it is not required that the order p be chosen minimally. Let π denote the projection of the fibre bundle $\pi : T^*X \setminus 0_X \to X$, i.e. the cotangent bundle with the zero section 0_X removed. To a differential operator D of order p we now define the p-symbol

$$\sigma^{(p)}(D) : \pi^*E \to \pi^*F$$

as follows (cf. [We80], p. 115):

For $0 \neq v \in T^*(X)$ and g a C^∞-function on X with $dg_x = v$ and $f(x) = e \in E_x$, put

$$\sigma^{(p)}(D)(x, v)e := D\left(\frac{i^p}{p!}(g - g(x))^p \cdot f\right)(x).$$

In order to describe the symbol locally, we form the matrix $\left(d_{ij}^{(p)}\right)$, in which $d_{ij}^{(p)}$ is the homogeneous component of d_{ij} of degree p. For $x \in X$ and $v = (v_1(v), \ldots, v_k(v)) \in T_x^*X$ we replace the $\frac{\partial}{\partial x_r}$ by iv_r in $\left(d_{ij}^{(p)}\right)$ and obtain a matrix $\left(\sigma_{ij}^{(p)}\right)$. This is independent of the chosen chart of the manifold, since the v_r (coordinate functions of T_x^*X) and the $\frac{\partial}{\partial x_r}$ (basis of T_xX) coincide. We therefore obtain the $\mathbb{C}$-linear map $\left(\sigma_{ij}^{(p)}(D)(x, v)\right)$. Obviously, the symbol depends on the order of the differential operator.

Definition: *A differential operator D of order p is called an **elliptic differential operator** if $\sigma^{(p)}(D) : \pi^*E \to \pi^*F$ is a bundle isomorphism of π^*E onto π^*F.*

Remark: For an elliptic differential operator the order is therefore well-defined.

Example: On $X = \mathbb{R}^k/L$, where $L \subset \mathbb{R}^k$ is a lattice, the Laplace operator $\Delta :=$ $\frac{\partial^2}{\partial x_1^2} + \cdots + \frac{\partial^2}{\partial x_k^2}$ is an elliptic operator of order two on functions, i.e. on sections of the trivial bundle $X \times \mathbb{C}$, for we have $\sigma^{(2)}(\Delta)(x,v) = -\left(v_1^2 + \cdots + v_k^2\right) \neq 0$ for $v \neq 0$.

For elliptic differential operators one has the following (cf. [We80]) important

Theorem: $\ker(D)$ *and* $\operatorname{coker}(D)$ *are finite dimensional.*

Remark: One obtains the statement about the cokernel from the (difficult) statement about the kernel, by considering the adjoint operator.

Definition: *The **index** of D is defined by*

$$\operatorname{ind}(D) = \dim_{\mathbb{C}} \ker(D) - \dim_{\mathbb{C}} \operatorname{coker}(D).$$

Now let $E_0, \ldots, E_m$ be complex vector bundles over X, and for $i = 0, \ldots, m-1$ let $D_i : \Gamma E_i \to \Gamma E_{i+1}$ be differential operators of degree p with $D_{i+1} D_i = 0$ and corresponding symbols $\sigma_i = \sigma^{(p)}(D_i)$.

Definition: *Such a complex*

$$D : \Gamma E_0 \xrightarrow{D_0} \Gamma E_1 \to \cdots \xrightarrow{D_{m-1}} \Gamma E_m$$

*is called an **elliptic complex** if the corresponding complex of symbols*

$$0 \to \pi^* E_0 \xrightarrow{\sigma_0} \cdots \xrightarrow{\sigma_{m-1}} \pi^* E_m \to 0$$

is an exact sequence of vector bundles.

Remark: If one considers again $\sigma_i^{(p)}(x,v) : E_{i,x} \to E_{i+1,x}$, then the complex is elliptic if and only if the complex of symbols is exact for all $x \in X$ and $v \in T_x^* X \setminus \{0\}$. For $m = 1$, an elliptic complex is precisely an elliptic differential operator.

Since the composition $D_i D_{i-1} = 0$, one can consider the cohomology groups $H^i = \ker(D_i)/\operatorname{im}(D_{i-1})$. From the assertion above for the case $m = 1$ now follows easily the

Theorem: *For an elliptic complex, all cohomology groups H^i are finite dimensional.*

We shall put $h^i := \dim_{\mathbb{C}} H^i$ when it is clear which elliptic complex is meant.

Definition: *The **index** of an elliptic complex D is defined by*

$$\operatorname{ind}(D) = \sum_{i=0}^{m} (-1)^i \cdot \dim_{\mathbb{C}} H^i = \sum_{i=0}^{m} (-1)^i \cdot h^i.$$

For $m = 1$, the index of an elliptic complex coincides with the index of the elliptic operator D_0.

5.2 The index of an elliptic complex

Atiyah-Singer Index Theorem: *Let X be a compact, oriented, differentiable manifold of dimension $2n$ and $D = (D_i : \Gamma E_i \to \Gamma E_{i+1})$ an elliptic complex ($i = 0, \cdots, m-1$), associated to the tangent bundle. Then the index of this complex is determined by the following formula (cf. [AtSi68I], [AtSe68II], [AtSi68III]):*

$$\operatorname{ind}(D) = (-1)^n \left(\left(\frac{1}{e(T^*X)} \sum_{i=0}^{m} (-1)^i \cdot \operatorname{ch}(E_i) \right) \operatorname{td}(TX \otimes \mathbb{C}) \right)[X]. \qquad \square$$

Remark: In this formula, td is the multiplicative sequence associated to the power series $Q(x) = x/(1 - e^{-x})$ (the Todd genus, cf. section 1.8). The alternating sum of the Chern characters is in fact divisible by the Euler class; this is a non-trivial fact. It is convenient for calculations to factor the Euler class $x_1 \cdots x_n$ formally out of the Todd class of $TX \otimes \mathbb{C}$. The resulting expression no longer makes sense topologically; for concrete calculations the Euler class always appears as a factor in the Chern characters and corrects this error. We therefore use the formula

$$\operatorname{ind}(D) = \left(\left(\sum_{i=0}^{m} (-1)^i \cdot \operatorname{ch}(E_i) \right) \prod_{j=1}^{n} \left(\frac{x_j}{1 - e^{-x_j}} \cdot \frac{1}{1 - e^{x_j}} \right) \right)[X].$$

This formula is, however, to be used with caution, since the right product is neither symmetric in the x_j^2 nor an invertible power series.

Remark: This theorem gives a purely topological expression for the differentiable invariant $\operatorname{ind}(D)$. In particular, the differential operators themselves make no appearance in the formula. Interpreted differently, one now knows that the complicated expression on the right side is integral.

The examples in the following three sections are complexes of order one.

5.3 The de Rham complex

Let X be a compact, differentiable manifold of dimension k and T its tangent bundle. The bundle of differential forms of degree i with complex-valued coefficients is $E_i = \Lambda^i(T^* \otimes \mathbb{C})$. The exterior derivative d yields an elliptic complex, $d = (d : \Gamma E_i \to \Gamma E_{i+1})$, whose symbol $\sigma_i(x, v)$ is the wedge-product $\omega \mapsto iv \wedge \omega$ (as linear map). The cohomology groups H^i will also be denoted by $H^i_{\text{de Rham}}$. By the theorem of de Rham, we have:

$$H^i_{\text{de Rham}} \cong H^i(X; \mathbb{C}).$$

This means that not only the index of the elliptic complex d, but indeed all the individual $h^i = \dim H^i_{\text{de Rham}}$ are topological invariants, which we also denote as Betti numbers b_i. Thus

$$\text{ind}\,(d) = \sum_{i=0}^{k} (-1)^i \cdot h^i = \sum_{i=0}^{k} (-1)^i \cdot b_i = e(X).$$

We now compare this result with the Atiyah-Singer index theorem. For this, let X now be $2n$-dimensional and oriented. As we have seen in section 1.5, we have:

$$\text{ch}\Big(\sum_{i=0}^{n} \big(\Lambda^i(T^* \otimes \mathbb{C})\big) \cdot y^i\Big) = \prod_{j=1}^{n} \big((1 + ye^{x_j})(1 + ye^{-x_j})\big).$$

For $y = -1$ this is exactly the sum in the Atiyah-Singer index formula. It cancels all factors there apart from the Euler class, and so gives correctly:

$$\text{ind}\,(d) = (e(TX))[X] = e(X).$$

5.4 The Dolbeault complex

Now let X be a complex n-dimensional manifold, T the holomorphic tangent bundle. We have

$$T_{\mathbb{R}}^* \otimes \mathbb{C} \cong T^* \oplus \overline{T^*}$$
$$\Rightarrow \quad \Lambda^i(T_{\mathbb{R}}^* \otimes \mathbb{C}) \cong \Lambda^i\big(T^* \oplus \overline{T^*}\big)$$
$$\cong \bigoplus_{p+q=i} \big(\Lambda^p T^* \otimes \Lambda^q \overline{T^*}\big).$$

With the notation $A^{p,q} := \Gamma\big(\Lambda^p T^* \otimes \Lambda^q \overline{T^*}\big)$, the exterior derivative $d : A^{p,q} \to A^{p+1,q} \oplus A^{p,q+1}$ splits into $d = \partial + \overline{\partial}$, with $\partial : A^{p,q} \to A^{p+1,q}$ and $\overline{\partial} : A^{p,q} \to A^{p,q+1}$. For fixed p, $\overline{\partial}$ yields an elliptic complex of differential operators. This is called the Dolbeault complex. Let $H^{p,q}$ be the q-th cohomology group of this complex, $h^{p,q} := \dim_{\mathbb{C}} H^{p,q}$. Hence $H^{p,0}$ is the vector space of holomorphic sections of $\Lambda^p T^*$, the bundle of holomorphic p-forms on X. For $p = 0$ these are the holomorphic functions on X, which must be locally constant since X is compact. Hence $h^{0,0}$ is the number of connected components of X. The index of $\overline{\partial}$ for fixed p is also denoted by χ^p :

$$\chi^p := \sum_{q=0}^{n} (-1)^q \cdot h^{p,q}.$$

In addition, $\chi := \chi^0$ is called the arithmetic genus of X.

We shall now calculate these χ^p by means of the Atiyah-Singer index theorem. We reduce the structure group of the holomorphic tangent bundle T to $U(n)$, hence can identify $\overline{T^*}$ with T. First we consider the case $p = 0$. The bundles appearing are then $\Lambda^q \overline{T^*} \cong \Lambda^q T$. From our earlier result (cf. section 1.5)

$$\sum_{i=0}^{n} \mathrm{ch}\,(\Lambda^q T) \cdot y^q = \prod_{j=1}^{n} (1 + y \cdot e^{x_j}),$$

we obtain, with $y = -1$, for the sum in the Atiyah-Singer index theorem the expression $\prod_{j=1}^{n} (1 - e^{x_j})$. Hence there remains in the formula only one of the two factors in the product:

$$\chi^0 = \prod_{j=1}^{n} \left(\frac{x_j}{1 - e^{-x_j}} \right)[X] = \mathrm{td}\,(T)[X] = \mathrm{td}\,(X).$$

For the first three dimensions we have as examples (cf. section 1.8):

$$n = 1: \quad \chi^0 = \frac{c_1}{2}\,[X] = \frac{e(X)}{2}, \qquad n = 2: \quad \chi^0 = \frac{c_1^2 + c_2}{12}\,[X],$$

$$n = 3: \quad \chi^0 = \frac{c_1 c_2}{24}\,[X].$$

All χ^0 are integral, being the index of an operator. One therefore obtains divisibility conditions on the Chern numbers. For $n = 1$, χ^0 is half the Euler number, hence $\chi^0 = 1 - g$ where g is the usual genus (number of handles) of the Riemann surface.

For general p we must determine the weights of the representation which leads to $\Lambda^p T^* \otimes \Lambda^q \overline{T^*}$, in order to be able to identify the Chern character. If we collect together the results for different p by putting

$$\chi_y := \sum_{p=0}^{n} \chi^p \cdot y^p,$$

we obtain the

Proposition: $\quad \chi_y = \prod_{j=1}^{n} \left((1 + y \cdot e^{-x_j}) \frac{x_j}{1 - e^{-x_j}} \right)[X].$

Proof: This follows easily from the formulas in section 1.5.

Remark: The power series $x(1 + y \cdot e^{-x})/(1 - e^{-x})$ can unfortunately not be taken as the characteristic power series for a (complex) genus, since it starts with $1 + y$ instead of 1. If we replace x by $x(1 + y)$, then we obtain by evaluation on a $2n$-dimensional manifold an additional factor $(1 + y)^n$, since for the evaluation only the homogeneous component of degree n in the x_j enters. We therefore obtain:

$$\chi_y = \prod_{j=1}^{n} \left(\left(1 + y \cdot e^{-x_j}\right) \frac{x_j}{1 - e^{-x_j}} \right) [X]$$

$$= \prod_{j=1}^{n} \left(\frac{1 + y \cdot e^{-x_j(1+y)}}{1 + y} \frac{x_j(1+y)}{1 - e^{-x_j(1+y)}} \right) [X]$$

$$= \prod_{j=1}^{n} \left(\frac{x_j\left(1 + y \cdot e^{-x_j(1+y)}\right)}{1 - e^{-x_j(1+y)}} \right) [X].$$

Hence χ_y is the genus belonging to the power series

$$Q(x) = \frac{x\left(1 + y \cdot e^{-x(1+y)}\right)}{1 - e^{-x(1+y)}}.$$

For three values of y, this χ_y-genus is an important invariant:

$y = 0$:

$$\chi_0 = \sum_{p=0}^{n} \chi^p \cdot 0^p = \chi^0 = \chi \qquad \text{(arithmetic genus)}.$$

$y = -1$:

$$\chi_{-1} = \sum_{p=0}^{n} \chi^p \cdot (-1)^p = \sum_{p=0}^{n} \left((-1)^p \cdot \sum_{q=0}^{n} (-1)^q \cdot h^{p,q} \right)$$

$$= \sum_{p,q=0}^{n} (-1)^{p+q} \cdot h^{p,q} = \operatorname{ind}\left(\overline{\partial}\right) = \operatorname{ind}(d) = e(X)$$

($\operatorname{ind}\left(\overline{\partial}\right) = \operatorname{ind}(d)$, since in the Atiyah-Singer index theorem the same bundles, namely $\Lambda^i T_{\mathbb{R}}^* \otimes \mathbb{C}$, appear).

The characteristic power series for the Euler number can therefore not obviously be normalized (for $y = -1$, the numerator and the denominator of $Q(x)$ both are zero). Nonetheless, an application of l'Hospital's rule yields $Q(x) = 1 + x$.

$y = 1$:   As characteristic power series remains

$$\frac{x\left(1 + e^{-2x}\right)}{1 - e^{-2x}} = \frac{x\left(e^x + e^{-x}\right)}{e^x - e^{-x}} = \frac{x}{\tanh(x)}.$$

As we already know, this even power series yields the signature of the manifold, so

$$\chi_1 = \sum_{p=0}^{n} \chi^p = \operatorname{sign}(X).$$

Remark: For Kähler manifolds the two formulas $\chi_{-1} = \sum_{p,q=0}^{n} (-1)^{p+q} \cdot h^{p,q} = e(X)$ and $\chi_1 = \sum_{p,q=0}^{n} (-1)^q \cdot h^{p,q} = \text{sign}(X)$ were already known earlier (cf. [Ho50]), since then $H^m \cong \bigoplus_{p+q=m} H^{p,q}$.

5.5 The signature as an index

For an n-dimensional oriented vector space V, provided with a scalar product, one can define a linear map (Hodge $*$-operator) $* : \Lambda^i V \to \Lambda^{n-i} V$. To do so, let $e_1, \ldots, e_n$ be an orthonormal basis of V compatible with the orientation. The mapping $*$ is then defined on a basis of $\Lambda^i V$ by

$$*(e_{j_1} \wedge \cdots \wedge e_{j_i}) = e_{j_{i+1}} \wedge \cdots \wedge e_{j_n},$$

where the indices $j_{i+1}, \ldots, j_n$ are ordered so that $e_{j_1} \wedge \cdots \wedge e_{j_n}$ gives the orientation of V. This definition does not depend on the choice of orthonormal basis. The scalar product on V induces a scalar product $\langle \cdot, \cdot \rangle$ on $\Lambda^i V$, with respect to which the $e_{j_1} \wedge \cdots \wedge e_{j_i}$ with $j_1 < \cdots < j_i$ form an orthonormal basis. Therefore for $\alpha, \beta \in \Lambda^i V$ and $\omega = e_1 \wedge \cdots \wedge e_n$ we have:

$$\alpha \wedge *\beta = \langle \alpha, \beta \rangle \cdot \omega.$$

Moreover, one sees easily that $** = (-1)^{i(n-i)} \cdot \text{id}$. For this reason we define a map τ on $\Lambda^i V$ by

$$\tau := (-1)^{i(i-1)/2+k} *$$

if $\dim V = 4k$. Then $\tau^2 = \text{id}$ holds.

We now compute the signature of a manifold as the index of an elliptic operator (cf. [AtSi68III]). Thus let X be a compact, oriented, differentiable, $4k$-dimensional manifold and let T be its tangent bundle provided with a Riemannian metric. This induces a metric on the bundles $\Lambda^i(T^* \otimes \mathbb{C})$. As above (consider local orthonormal bases) the $*$- and τ-operators can be defined on the exterior powers of the cotangent space and also on the sections $A^i = \Gamma(\Lambda^i(T^* \otimes \mathbb{C}))$. Then for $\alpha, \beta \in A^i$ we have

$$\langle \alpha, \beta \rangle = \int_X \alpha \wedge *\overline{\beta}.$$

Since $\tau^2 = \text{id}$, τ has the eigenvalues $+1$ and -1. We can extend τ linearly to $\Lambda(T^* \otimes \mathbb{C})$ and obtain the sub-bundles E_+ with eigenvalue $+1$ and E_- with eigenvalue -1. For $\alpha \in \Lambda^i(T^* \otimes \mathbb{C})$ we have $\alpha + \tau\alpha \in E_+$, $\alpha - \tau\alpha \in E_-$ and $\alpha = \frac{1}{2}((\alpha + \tau\alpha) + (\alpha - \tau\alpha))$, hence $\Lambda(T^* \otimes \mathbb{C}) = E_+ \oplus E_-$. The maps $\alpha \mapsto \alpha \pm \tau\alpha$

are injective for $\alpha \in \Lambda^i(T^* \otimes \mathbb{C})$ with $i < 2k$, and because of the dimension shift the rank of $E_\pm$ is therefore greater than or equal to $\sum_{i=0}^{2k-1} \binom{4k}{i}$. In the critical case $i = 2k$ one can easily show that the rank of the bundle $E_\pm \cap \Lambda^{2k}(T^* \otimes \mathbb{C})$ is precisely $\frac{1}{2}\binom{4k}{2k}$. Therefore we have

$$\mathrm{rk}_{\mathbb{C}} E_+ = \mathrm{rk}_{\mathbb{C}} E_- = 2^{4k-1}.$$

As usual, $d : A^i \to A^{i+1}$ is the exterior derivative. We define $d^* : A^i \to A^{i-1}$ by $d^* = - *d* = -\tau d\tau$. The Hodge $*$-operator is a differential operator of order zero and d has order one. For $\alpha \in A^i$ and $\beta \in A^{i+1}$ we have:

$$\begin{aligned}
\langle d\alpha, \beta \rangle &= \int_X d\alpha \wedge *\overline{\beta} \\
&= \int_X (-1)^{i+1} \alpha \wedge d* \overline{\beta} + \int_X d(\alpha \wedge *\overline{\beta}) \\
&= \int_X (-1)^{i+1+i(4k-i)+1} \alpha \wedge *(- *d*)\overline{\beta} \\
&= \langle \alpha, d^*\beta \rangle.
\end{aligned}$$

Hence d^* is adjoint to d, and therefore the differential operator $d + d^*$ is self-adjoint, so its index is zero. Since

$$\begin{aligned}
\tau(d + d^*) &= \tau d - \tau\tau d\tau = \tau d - d\tau = -(d\tau - \tau d) \\
&= -(d\tau + d^*\tau) = -(d + d^*)\tau,
\end{aligned}$$

one can consider $d + d^* : \Gamma E_+ \to \Gamma E_-$.

In order to determine the signature of X, we must decompose $H^{2k}_{\mathrm{de\ Rham}} = H^{2k}(X;\mathbb{C})$ into a positive and a negative definite part with respect to the intersection form. Now in each de Rham cohomology class there is precisely one harmonic form as representative. These are the elements of $\Gamma(\Lambda(T^* \otimes \mathbb{C}))$ which lie in the kernel of $(d + d^*)^2 = \Delta$ (Laplace operator) and therefore (since $d + d^*$ is self-adjoint) are in the kernel of $d + d^*$. Let H^{2k}_+ (resp. H^{2k}_-) be those subspaces of $H^{2k}(X;\mathbb{C})$ which correspond to the harmonic forms α with $\tau\alpha = \alpha$ (resp. $\tau\alpha = -\alpha$). Therefore $H^{2k}(X;\mathbb{C}) = H^{2k}_+ \oplus H^{2k}_-$ is a decomposition into eigenspaces with respect to τ, which induces a decomposition of $H^{2k}(X;\mathbb{R})$. For a real harmonic form $\alpha \in H^{2k}_+$, $\alpha \neq 0$, we have:

$$0 < \langle \alpha, \alpha \rangle = \int_X \alpha \wedge *\alpha = \int_X \alpha \wedge \tau\alpha = \int_X \alpha \wedge \alpha.$$

For real $\alpha \in H^{2k}_-$, $\alpha \neq 0$, holds correspondingly:

$$0 > -\langle \alpha, \alpha \rangle = - \int_X \alpha \wedge *\alpha = - \int_X \alpha \wedge \tau\alpha = \int_X \alpha \wedge \alpha.$$

Since τ is self-adjoint on H^{2k}, we also have $\int_X \alpha \wedge \beta = 0$ if $\alpha \in H^{2k}_+$ and $\beta \in H^{2k}_-$. Therefore we have decomposed the intersection form on H^{2k} into a positive definite and a negative definite part, and thus

$$\operatorname{sign}(X) = \dim_{\mathbb{C}} H^{2k}_+ - \dim_{\mathbb{C}} H^{2k}_-.$$

Now we consider the operator $d + d^* : \Gamma E_+ \to \Gamma E_-$. This is elliptic, since $(d + d^*)^2 = \Delta$ is the Laplace operator. Its kernel consists of the harmonic forms for the eigenvalue 1 of τ. The adjoint operator is $d + d^* : \Gamma E_- \to \Gamma E_+$, whose kernel consists of the harmonic forms for the eigenvalue -1. In the expression

$$\operatorname{ind}(d + d^*) = \dim \ker(d + d^*) - \dim \operatorname{coker}(d + d^*)$$
$$= \dim \ker(d + d^* : \Gamma E_+ \to \Gamma E_-) - \dim \ker(d + d^* : \Gamma E_- \to \Gamma E_+),$$

the components of harmonic forms away from the middle-dimensional cohomology have the same dimension and so cancel. There only remains

$$\operatorname{ind}(d + d^*) = \dim H^{2k}_+ - \dim H^{2k}_-.$$

Hence the signature of X is precisely the index of $d + d^*$.

Again, we compare the result with the Atiyah-Singer index theorem. For $\operatorname{ch}(E_+) - \operatorname{ch}(E_-)$ one can show (cf. [AtHi59a]):

$$\operatorname{ch}(E_+) - \operatorname{ch}(E_-) = \prod_{j=1}^{2k} \left(e^{x_j} - e^{-x_j}\right).$$

We recast the product term in the Atiyah-Singer index theorem slightly:

$$1 - e^{-x_j} = e^{-x_j/2} \cdot \left(e^{x_j/2} - e^{-x_j/2}\right),$$

$$1 - e^{x_j} = e^{x_j/2} \cdot \left(e^{-x_j/2} - e^{x_j/2}\right)$$

$$\Rightarrow \quad \prod_{j=1}^{2k} \left(\frac{x_j}{1 - e^{-x_j}} \cdot \frac{1}{1 - e^{x_j}}\right) = \prod_{j=1}^{2k} \frac{x_j}{\left(e^{x_j/2} - e^{-x_j/2}\right)^2}$$

$$\Rightarrow \quad \operatorname{sign}(X) = \left(\prod_{j=1}^{2k} \left(e^{x_j} - e^{-x_j}\right) \prod_{j=1}^{2k} \frac{x_j}{\left(e^{x_j/2} - e^{-x_j/2}\right)^2}\right)[X]$$

$$= \left(\prod_{j=1}^{2k} \frac{x_j\left(e^{x_j/2} + e^{-x_j/2}\right)}{\left(e^{x_j/2} - e^{-x_j/2}\right)}\right)[X]$$

$$= \left(\prod_{j=1}^{2k} \frac{x_j}{\tanh(x_j/2)}\right)[X] = \frac{1}{2^{2k}}\left(\prod_{j=1}^{2k} \frac{2x_j}{\tanh(x_j)}\right)[X]$$

$$= \left(\prod_{j=1}^{2k} \frac{x_j}{\tanh(x_j)}\right)[X].$$

One obtains the next-to-last equation by replacing x_j with $2x_j$. Since we evaluate over X, only the homogeneous components of degree $2k$ in the x_j must agree, and therefore one must divide by 2^{2k}.

5.6 The equivariant index

Now let X be a compact, complex manifold of complex dimension n and as before let $D = (D_i : \Gamma E_i \to \Gamma E_{i+1})$ be an elliptic complex. We want to generalize the Atiyah-Singer index theorem to the case in which a compact topological group G acts on X by holomorphic maps (cf. [AtBo67], [AtSi68III]). Let $X^g = \{x \in X \mid gx = x\}$ denote the fixed point set under the operation of a fixed group element $g \in G$. This is a complex submanifold of X, which is not necessarily connected. We therefore decompose $X^g = \bigcup X^g_\nu$ into its connected components. The X^g_ν are connected submanifolds of X of possibly different dimensions. In addition, assume that G acts on the elliptic complex, i.e. G acts on the bundles E_i (e.g. if all bundles E_i are associated to the tangent bundle of X) and this action commutes with the differential operators D_i. Then G also acts on the cohomology groups H^i of the complex. Hence, for each element $g \in G$ the trace $\operatorname{tr}(g, H^i)$ of the action of g on H^i is defined.

Definition: *In the above situation the **equivariant index** $\operatorname{ind}(g, D)$ of D is defined as*

$$\operatorname{ind}(g, D) := \sum_{i=0}^{m} (-1)^i \cdot \operatorname{tr}(g, H^i).$$

Example: $\operatorname{ind}(D) = \operatorname{ind}(\mathrm{id}, D)$.

Example: The index of the Dolbeault complex is $\chi^p(X) = \sum_{q=0}^{n} (-1)^q \cdot \dim H^{p,q}$. The spaces $H^{p,q}$ are derived from the complex structure of the manifold X. If g is a holomorphic automorphism of X, then g acts on all $H^{p,q}$. The equivariant index is then:

$$\chi^p(g, X) = \sum_{q=0}^{n} (-1)^q \cdot \operatorname{tr}(g, H^{p,q}).$$

We now want to indicate how one computes the equivariant index as a sum of contributions $a(X^g_\nu)$ corresponding to the fixed point components X^g_ν. Let $Y = X^g_\nu$ be one of the fixed point components of X for an element $g \in G$. For a point $p \in Y$, g acts linearly on the tangent space $T_p X$. There exists a Hermitian metric on $TX|_Y$ so that g acts unitarily on $TX|_Y$. Therefore $T_p X$ decomposes into the direct sum of eigenspaces $N_{p,\lambda}$ for eigenvalues λ of modulus one. Since G acts continuously on X, one further obtains an eigenbundle N_λ over Y. Indeed, N_1 is precisely the tangent bundle TY to Y. Under variation of the point p in Y, the eigenvalues cannot change since they depend only on the isomorphism type of the representation on T_pX of the subgroup generated by g, and the representation ring of this subgroup is discrete. With $d_\lambda = \operatorname{rk} N_\lambda$ we therefore have:

$$TX|_Y = \bigoplus_\lambda N_\lambda, \qquad c(N_\lambda) = \prod_{i=1}^{d_\lambda} \left(1 + x_i^{(\lambda)}\right)$$

$$\Rightarrow \quad c(TX|_Y) = \prod_\lambda \prod_{i=1}^{d_\lambda} \left(1 + x_i^{(\lambda)}\right) = \prod_{i=1}^{n} (1 + x_i).$$

This means that each of the formal roots x_i can be identified with one of the formal roots $x_i^{(\lambda)}$ belonging to a definite eigenvalue λ. The recipe for the calculation of $a(Y)$ is now as follows:

We consider the original index formula

$$\operatorname{ind}(D) = \left(\left(\sum_{i=0}^{m} (-1)^i \cdot \operatorname{ch}(E_i)\right) c_n(X) \prod_{i=1}^{n} \left(\frac{1}{1 - e^{-x_i}} \cdot \frac{1}{1 - e^{x_i}}\right)\right)[X].$$

In this formula, we replace X by Y, e^{x_i} by $\lambda^{-1} \cdot e^{x_i}$ and e^{-x_i} by $\lambda \cdot e^{-x_i}$ (where x_i belongs to the eigenvalue λ). We apply the same process to the terms $\operatorname{ch}(E_i)$. This can obviously be done if the E_i are associated to the tangent bundle of X. For $\lambda = 1$, the term $1 - \lambda^{-1} \cdot e^{x_i}$ is not invertible; but those x_i, which belong to the eigenvalue 1, originate from the tangent bundle to Y and so cancel with the factor $c_n(Y)$.

Example: Applying this recipe to the χ_y-genus, we obtain from

$$\sum_{p=0}^{n} \chi^p(X) y^p = \chi_y(X) = \left(\prod_{i=1}^{n} (1 + y \cdot e^{-x_i}) \frac{x_i}{1 - e^{-x_i}}\right)[X]$$

the equivariant formula

$$\sum_{p=0}^{n} \chi^p(g, X) \cdot y^p = \chi_y(g, X) = \sum_\nu a(X_\nu^g),$$

where we have for each fixed point component $Y = X_\nu^g$:

$$a(Y) = \left(\prod_{\lambda \neq 1} \prod_{i=1}^{d_\lambda} \left(\frac{1 + y \cdot \lambda e^{-x_i^{(\lambda)}}}{1 - \lambda e^{-x_i^{(\lambda)}}}\right) \cdot \prod_{i=1}^{d_1} \left(1 + y e^{-x_i^{(1)}}\right) \frac{x_i^{(1)}}{1 - e^{-x_i^{(1)}}}\right)[Y].$$

If we assume in addition that such a fixed point component consists of a single point, i.e. $\dim Y = 0$, then all eigenvalues are distinct from 1; and for the evaluation on Y only the zero-dimensional component is relevant, hence

$$a(Y) = \prod_\lambda \prod_{i=1}^{d_\lambda} \frac{1 + \lambda y}{1 - \lambda}.$$

Since the denominator is exactly the characteristic polynomial evaluated at 1, it does not vanish (1 is not an eigenvalue).

For the equivariant Euler number of X we have in the general case

$$e(g, X) = \chi_{-1}(g, X) = \sum_{\nu} a(X_\nu^g)_{y=-1}$$

$$= \sum_{\nu} \Big(\prod_{i=1}^{d_1} x_i^{(1)} \Big) [X_\nu^g] = \sum_{\nu} e(X_\nu^g),$$

and in the isolated fixed points case

$$a(Y)_{y=-1} = \prod_{\lambda} \prod_{i=1}^{d_\lambda} \frac{1-\lambda}{1-\lambda} = 1,$$

what we expected from the above since it is the Euler number of the point Y. For the equivariant signature ($n = \dim_{\mathbb{C}} X$) we have (again assuming isolated fixed points)

$$a(Y)_{y=1} = \prod_{\lambda} \prod_{i=1}^{d_\lambda} \frac{1+\lambda}{1-\lambda} = (-1)^n \frac{\text{char. polynomial } (-1)}{\text{char. polynomial } (1)} .$$

5.7 The equivariant χ_y-genus for S^1-actions

Now let $G = S^1$ and let $q \in S^1$ be a topological generator, i.e. the subgroup of S^1 generated by q is dense in S^1. Every torus has a topological generator. Then we have

$$X^q = X^{S^1} = \{ x \in X \mid gx = x \text{ for all } g \in S^1 \}.$$

In this situation, the whole group S^1 acts on the restriction of the tangent bundle of X to X^q, not merely a group element. The irreducible representations of S^1 are all one-dimensional, elements $g \in S^1$ acting as multiplication by g^k for fixed $k \in \mathbb{Z}$. If S^1 acts on a vector space, then one also writes

$$V = \sum_{k=-\infty}^{\infty} q^k V_k$$

which means that V is a direct sum $V = \bigoplus_{k=-\infty}^{\infty} V_k$ and q acts on V_k as multiplication by q^k. As in section 5.6, we obtain a splitting of $TX|_Y$ (where $Y = X_\nu^{S^1}$ is a connected component of X^{S^1}):

$$TX|_Y = \sum_{k=-\infty}^{\infty} q^k N_k.$$

The eigenvalues λ of q are now all integral powers q^k of q.

The equivariant χ_y-genus is then ($d_k = \mathrm{rk}\, N_k$)

$$\chi_y(q, X) = \sum_\nu a\big(X_\nu^{S^1}\big)$$

with

$$a\big(X_\nu^{S^1}\big) = \left(\prod_{i=1}^{d_0} x_i^{(1)} \frac{1 + y \cdot e^{-x_i^{(1)}}}{1 - e^{-x_i^{(1)}}} \cdot \left(\prod_{k \neq 0} \prod_{j=1}^{d_k} \frac{1 + y \cdot q^k e^{-x_j^{(q^k)}}}{1 - q^k e^{-x_j^{(q^k)}}}\right)\right)[X_\nu^{S^1}].$$

Because $\chi_y(q, X)$ is a finite (alternating) sum of traces of the action of q on finite dimensional vector spaces for which q has only eigenvalues of the form q^k, $\chi_y(q, X)$ is a finite Laurent series in q. On the other hand, the expression on the right side is a rational function of q. Since this holds for all $q \in S^1$, this is in fact an identity in the indeterminate q.

We therefore can consider the limit $q \to 0$. For positive k the factors on the right side all tend to 1, while for negative k we multiply by q^{-k} and see that the factors converge to $-y$. We therefore have

$$\begin{aligned}
a\big(X_\nu^{S^1}\big)_{q=0} &= \left(\prod_{i=1}^{d_0} x_i^{(1)} \frac{1 + y \cdot e^{-x_i^{(1)}}}{1 - e^{-x_i^{(1)}}} \cdot \left(\prod_{k<0} \prod_{j=1}^{d_k} (-y)\right)\right)[X_\nu^{S^1}] \\
&= \left(\prod_{i=1}^{d_0} x_i^{(1)} \frac{1 + y \cdot e^{-x_i^{(1)}}}{1 - e^{-x_i^{(1)}}}\right)[X_\nu^{S^1}] \cdot \prod_{k<0} (-y)^{d_k} \\
&= \chi_y\big(X_\nu^{S^1}\big) \cdot (-y)^{\sum_{k<0} d_k}.
\end{aligned}$$

Now for the limit $q \to \infty$. This time we obtain the factor $-y$ for positive k, and the factor 1 for negative k, so that

$$a\big(X_\nu^{S^1}\big)_{q=\infty} = \chi_y\big(X_\nu^{S^1}\big) \cdot (-y)^{\sum_{k>0} d_k}.$$

Therefore $\chi_y(q, X)$ has a value for $q = 0$ as well as for $q = \infty$. Since it is a finite Laurent series in q, it must be constant in q and

$$\chi_y(q, X) \equiv \chi_y(\mathrm{id}, X) = \chi_y(X).$$

In particular, the values of $\chi_y(q, X)$ and so the sum of the $a\big(X_\nu^{S^1}\big)$ coincide for $q \to 0$ and for $q \to \infty$.

These considerations are essentially due to Lusztig [Lu71], Kosniowski [Ko70], and Atiyah and Hirzebruch [AtHi70].

Remark: This is also clear directly for a Kähler manifold. The spaces $H^{p,q}$ are then direct summands of the cohomology. Because S^1 is connected, each element acts homotopically to the identity, hence acts trivially on the cohomology and so on $H^{p,q}$.

Examples:

$y = -1$:

$$e(X) = \sum_\nu e\left(X_\nu^{S^1}\right) \cdot 1^{\sum_{k>0} d_k^{(\nu)}}$$
$$= \sum_\nu e\left(X_\nu^{S^1}\right)$$

$y = 1$:

$$\operatorname{sign}(X) = \sum_\nu \operatorname{sign}\left(X_\nu^{S^1}\right) \cdot (-1)^{\sum_{k>0} d_k^{(\nu)}}$$
$$= \sum_\nu \operatorname{sign}\left(X_\nu^{S^1}\right) \cdot (-1)^{\sum_{k<0} d_k^{(\nu)}}$$

Now assume that all fixed components $X_\nu^{S^1}$ are zero-dimensional. Then

$$\chi_y(X) = \sum_\nu (-y)^{\sum_{k<0} d_k^{(\nu)}} = \sum_\nu (-y)^{\sum_{k>0} d_k^{(\nu)}}$$
$$= \sum_\nu (-y)^{n - \sum_{k<0} d_k^{(\nu)}}$$

because the eigenvalue 1 ($k = 0$) does not appear. Therefore we have

$$\chi_y(X) = (-y)^n \chi_{y^{-1}}(X) = (-1)^n y^n \cdot \chi_{y^{-1}}(X).$$

This relation means that the polynomial $\chi_y(X)$ is symmetric in the coefficients of the y-powers (up to sign). Of course we could have derived this also from the well-known duality $h^{p,q} = h^{n-p,n-q}$:

$$\chi^p = \sum_{q=0}^n (-1)^q h^{p,q} = (-1)^n \sum_{q=0}^n (-1)^{n-q} h^{n-p,n-q}$$
$$= (-1)^n \chi^{n-p}.$$

Example: The circle S^1 acts on $P_2(\mathbb{C})$ by

$$(z_0 : z_1 : z_2) \overset{q}{\mapsto} \left(z_0 : q \cdot z_1 : q^2 \cdot z_2\right).$$

This action has three fixed points, namely

$$P_0 = (1 : 0 : 0), \ P_1 = (0 : 1 : 0), \ P_2 = (0 : 0 : 1).$$

In affine coordinates S^1 acts at

$$
\begin{aligned}
P_0 &\quad \text{by} \quad (z_1, z_2) \overset{q}{\mapsto} \left(q \cdot z_1, q^2 \cdot z_2\right) \\
P_1 &\quad \text{by} \quad (z_0, z_2) \overset{q}{\mapsto} \left(q^{-1} \cdot z_0, q \cdot z_2\right) \\
P_2 &\quad \text{by} \quad (z_0, z_1) \overset{q}{\mapsto} \left(q^{-2} \cdot z_0, q^{-1} \cdot z_1\right).
\end{aligned}
$$

Hence we have

$$
\begin{aligned}
\chi_y(P_2(\mathbb{C})) &= \chi_y(\mathrm{id}, P_2(\mathbb{C})) \\
&= \chi_y(P_2)(-y)^2 + \chi_y(P_1) \cdot (-y) + \chi_y(P_0) \\
&= \chi_y(P_2) + \chi_y(P_1) \cdot (-y) + \chi_y(P_0) \cdot (-y)^2 \\
&= y^2 - y + 1.
\end{aligned}
$$

In the same way one obtains for $P_n(\mathbb{C})$ the formula

$$
\chi_y(P_n(\mathbb{C})) = \sum_{i=0}^{n} (-y)^i,
$$

which we could also obtain from consideration of the Dolbeault complex.

5.8 The equivariant signature for S^1-actions

We can only define the χ_y-genus for complex manifolds. But for several values of y it is already defined for differentiable manifolds. These are for $y = -1$ the Euler number (not a genus) and for $y = 1$ the signature. For the Euler number we have already seen that its equivariant analogue is simply the sum of the Euler numbers of the fixed components. This also holds for differentiable manifolds. What happens for the signature?

Now let X be a $2n$-dimensional compact, oriented, differentiable manifold (up to now we have only considered the signature for even n). If we want to apply the equivariant Atiyah-Singer index theorem, we must again consider the decomposition of the bundle $TX|_{X_\nu^{S^1}}$ with respect to the action of a topological generator q of S^1. The irreducible representations of S^1 on real vector spaces are obtained from the complex representations by identification of $\mathbb{R}^2$ with $\mathbb{C}$. The multiplication with q^k then corresponds to a rotation in $\mathbb{R}^2$ through the k-fold angle of q. By suitable identification one can always take $k \geqslant 0$. We therefore obtain a splitting

$$
TX|_{X_\nu^{S^1}} = \sum_{k \geqslant 0} q^k N_k = TX_\nu^{S^1} \oplus \sum_{k > 0} q^k N_k.
$$

The bundles N_k are real bundles, which for $k > 0$ obtain a complex structure through the identification with a complex bundle. This identification yields a unique orientation on N_k, in view of the condition $k > 0$. Since the fixed component $X_\nu^{S^1}$ is orientable, $N_0 = TX_\nu^{S^1}$ can now be oriented so that all orientations taken together yield the orientation of X. In particular, according to this rule $X_\nu^{S^1}$ must sometimes be counted negatively when it is a point.

Having made this splitting, the equivariant index theorem yields:

$$\operatorname{sign}(q, X) = \sum_\nu a(X_\nu^{S^1}),$$

$$a(X_\nu^{S^1}) = \left(\prod_{i=1}^{d_0} x_i^{(1)} \frac{1 + e^{-x_i^{(1)}}}{1 - e^{-x_i^{(1)}}} \cdot \left(\prod_{k>0} \prod_{i=1}^{d_k} \frac{1 + q^k e^{-x_i^{(q^k)}}}{1 - q^k e^{-x_i^{(q^k)}}} \right) \right) [X_\nu^{S^1}].$$

Since the equivariant signature $\operatorname{sign}(q, X)$ does not depend on q (S^1 is connected, so acts trivially on cohomology), we can put $q = 0$ and have

$$a(X_\nu^{S^1}) = \left(\prod_{i=1}^{d_0} x_i^{(1)} \frac{1 + e^{-x_i^{(1)}}}{1 - e^{-x_i^{(1)}}} \right) [X_\nu^{S^1}] = \left(\prod_{i=1}^{d_0} x_i^{(1)} \frac{e^{x_i^{(1)}/2} + e^{-x_i^{(1)}/2}}{e^{x_i^{(1)}/2} - e^{-x_i^{(1)}/2}} \right) [X_\nu^{S^1}]$$

$$= \left(\prod_{i=1}^{d_0} \frac{x_i^{(1)}}{\tanh(x_i^{(1)}/2)} \right) [X_\nu^{S^1}] = \frac{1}{2^{d_0}} \left(\prod_{i=1}^{d_0} \frac{2x_i^{(1)}}{\tanh(x_i^{(1)})} \right) [X_\nu^{S^1}]$$

$$= \operatorname{sign}(X_\nu^{S^1}).$$

The next to last equation is obtained by replacing $x_i^{(1)}$ by $2x_i^{(1)}$ and correcting the error by division by 2^{d_0}.

We again assume that S^1 acts with isolated fixed points. Then the signature is

$$\operatorname{sign}(X) = \operatorname{sign}(q, X) = \sum_\nu \pm \prod_{k>0} \prod_{i=1}^{d_k^{(\nu)}} \frac{1 + q^k}{1 - q^k} = \sum_\nu (\pm 1).$$

The last equation holds again, since we know that the sum does not depend on q and we can therefore put $q = 0$. Here a fixed point receives the sign plus if $T_p X = \bigoplus_{k>0} N_k$ holds with the correct orientation, otherwise the sign minus.

Example: The circle S^1 acts on $S^{2n} \subset \mathbb{C}^n \times \mathbb{R}$ (coordinates $(z_1, \ldots, z_n, x)$) by rotation around the axis through the poles ($z_j \mapsto q \cdot z_j$ for $q \in S^1$ and $1 \leq j \leq n$). The fixed points are the north and south poles. The Euler number is $1 + 1 = 2$ and the signature is $1 - 1 = 0$ (the direction of rotation is once compatible with the orientation and once not).

6 Twisted operators and genera

6.1 Motivation for elliptic genera after Ed Witten

Let X again be a compact, oriented and differentiable manifold of dimension $4k$. The free loop space of X is the infinite dimensional manifold

$$\mathcal{L}X = \{g : S^1 \to X \mid g \text{ differentiable}\}.$$

There is a canonical action of the circle $S^1 \cong \mathbb{R}/\mathbb{Z}$ on $\mathcal{L}X$. If $g \in \mathcal{L}X$ is a loop, i.e. $g(x) \in X$ for $x \in S^1$, then the action of $t \in S^1$ on g is defined by $t(g)(x) := g(x - t)$. The fixed point set $(\mathcal{L}X)^{S^1}$ under this action is the manifold X itself, considered as the submanifold of constant loops in $\mathcal{L}X$. The tangent space at a loop $g \in \mathcal{L}X$ is $T_g(\mathcal{L}X) \cong \Gamma(g^*(TX))$, where as usual Γ denotes the sections of a bundle. For a constant loop $g \equiv p \in X$ the bundle $g^*(TX)$ is isomorphic to $S^1 \times T_pX$, so that

$$T_p(\mathcal{L}X) \cong \Gamma\big(S^1 \times T_pX\big) = \mathcal{L}(T_pX).$$

This decomposes into eigenspaces with respect to the S^1-action. Since $s \in \mathcal{L}(T_pX)$ is a mapping of $\mathbb{R}/\mathbb{Z}$ into a vector space V, s has a Fourier expansion:

$$s(x) = \frac{a_0}{2} + \sum_{n>0} \left(a_n \cos\left(2\pi n x\right) + b_n \sin\left(2\pi n x\right)\right) \in \mathcal{L}(T_pX).$$

Here the coefficients a_n, b_n are elements of T_pX. The circle $S^1 \cong \mathbb{R}/\mathbb{Z}$ acts as follows:

$$t(s)(x) = \frac{a_0}{2} + \sum_{n>0} \left(a_n \cos\left(2\pi n(x - t)\right) + b_n \sin\left(2\pi n(x - t)\right)\right) \in \mathcal{L}(T_pX).$$

Moreover, we have

$$
\begin{aligned}
& a_n \cos\left(2\pi n(x - t)\right) + b_n \sin\left(2\pi n(x - t)\right) \\
={}& a_n \cos\left(2\pi n x\right) \cdot \cos\left(2\pi n t\right) + a_n \sin\left(2\pi n x\right) \cdot \sin\left(2\pi n t\right) \\
& + b_n \sin\left(2\pi n x\right) \cdot \cos\left(2\pi n t\right) - b_n \cos\left(2\pi n x\right) \cdot \sin\left(2\pi n t\right) \\
={}& \widetilde{a}_n \cos\left(2\pi n x\right) + \widetilde{b}_n \sin\left(2\pi n x\right),
\end{aligned}
$$

where

$$\widetilde{a}_n + i\widetilde{b}_n = (a_n + ib_n) \cdot e^{2\pi int} = (a_n + ib_n)q^n \in T_pX \otimes \mathbb{C}$$

with $q = e^{2\pi it}$. We shall denote the bundle $TX \otimes \mathbb{C}$ by $T_{\mathbb{C}}$. Hence we have verified that

$$T(\mathcal{L}X)|_X = TX \oplus \sum_{n>0} q^n T_{\mathbb{C}}.$$

So we have bundles of finite rank, but infinitely many of them. It is probably possible to define directly an equivariant signature $\mathrm{sign}\,(q, \mathcal{L}X)$, under certain assumptions (such as orientability of the loop space). We shall achieve this indirectly: we apply the Atiyah-Singer index theorem and define the result to be the equivariant signature. With

$$p(X) = 1 + p_1 + \cdots + p_k = (1 + x_1^2) \cdots (1 + x_{2k}^2),$$
$$c(T_{\mathbb{C}}) = (1 + x_1) \cdots (1 + x_{2k})(1 - x_1) \cdots (1 - x_{2k}),$$

our recipe in section 5.8 applied to the loop space gives the

Definition: *The equivariant signature of the loop space is*

$$\mathrm{sign}\,(q, \mathcal{L}X) := \prod_{i=1}^{2k} \left(x_i \frac{1 + e^{-x_i}}{1 - e^{-x_i}} \cdot \prod_{n=1}^{\infty} \left(\frac{(1 + q^n e^{-x_i})(1 + q^n e^{x_i})}{(1 - q^n e^{-x_i})(1 - q^n e^{x_i})} \right) \right) [X].$$

This power series is symmetric in the x_i^2, and after evaluation on X is a power series in q with rational coefficients.

Definition: *Let X be a differentiable manifold of dimension $2k$ and W a complex vector bundle over X. Then the **signature** of X **with values in the vector bundle** W is defined as*

$$\mathrm{sign}\,(X, W) := \left(\prod_{i=1}^{k} \frac{x_i \cdot (1 + e^{-x_i})}{1 - e^{-x_i}} \cdot \mathrm{ch}\,(W) \right) [X].$$

Remark: This is the index of an elliptic operator, from which follows the integrality of $\mathrm{sign}\,(X, W)$: Recall the signature operator $d + d^* : \Gamma E_+ \to \Gamma E_-$, with the eigenspace bundles $E_+ = (\Lambda^* T \otimes \mathbb{C})^+$ and $E_- = (\Lambda^* T \otimes \mathbb{C})^-$ of the Hodge $*$-operator (on $\Lambda^* T \otimes \mathbb{C}$). By means of a connection on W, the signature operator can be extended to a twisted operator $\Gamma(E_+ \otimes W) \to \Gamma(E_- \otimes W)$ whose index is $\mathrm{sign}\,(X, W)$ (cf. [Pa65], IV, §9).

In order that the power series under the product begin with 1 (stable notation), we replace x_i by $2x_i$ and correct the error by dividing the factors by 2. If W has rank r, total Chern class $c(W) = (1 + y_1) \cdots (1 + y_r)$ and Chern character $\mathrm{ch}\,(W) = e^{y_1} + \cdots + e^{y_r}$, then the passage from y_j to $2y_j$ corresponds to the topological construction of a bundle $\Psi_2(W)$ (Adams operation), i.e. we have

$$\operatorname{ch}(\Psi_2(W)) = e^{2y_1} + \cdots + e^{2y_r}.$$

Hence

$$\operatorname{sign}(X, W) = \left(\prod_{i=1}^{k} \frac{x_i}{\tanh(x_i)} \ \operatorname{ch}(\Psi_2(W))\right)[X]$$

is stable and integral, and the first factor yields the L-class of the tangent bundle. All powers of two in the coefficients of the exponential function are canceled by the replacement of y with $2y$. The Adams operation therefore yields the vanishing of the powers of two in the denominator of the twisted signature; the absence of the other primes in the denominator remains an interesting matter.

Example: For X a sphere S^{2k} and an arbitrary vector bundle W over S^{2k} we have $\operatorname{sign}(S^{2k}, W) = \operatorname{ch}(\Psi_2(W))[S^{2k}] \in \mathbb{Z}$, since $p(S^{2k}) = 1$. Further,

$$\operatorname{ch}(\Psi_2(W))[S^{2k}] = \left(\sum_{i=1}^{r} \frac{(2y_i)^k}{k!}\right)[S^{2k}].$$

This sum can be expressed in the Chern classes of W. Since the cohomology of the sphere vanishes away from dimension $2k$, it suffices to compute the contribution of $c_k(W)$. This is (consider the multiplicative sequence for $Q(x) = 1 + x^k$ and the Cauchy lemma in section 1.8)

$$\left(\sum_{i=1}^{r} \frac{(2y_i^k)}{k!}\right)[S^{2k}] = (-1)^{k+1} \frac{2^k \cdot k}{k!} c_k(W)[S^{2k}].$$

Hence for each vector bundle W over S^{2k}

$$2^k \cdot c_k(W) \equiv 0 \ ((k-1)!).$$

At the end of section 6.2 we shall see that even $c_k(W) \equiv 0 \ ((k-1)!)$ holds.

Making use of the previous definitions $\Lambda_t T = \sum_{i=0}^{\infty} \Lambda^i T \cdot t^i$ and $S_t T = \sum_{i=0}^{\infty} S^i T \cdot t^i$, we now represent the equivariant signature of the loop space of a $4k$-dimensional manifold as an expression in twisted signatures:

Theorem: *We have*

$$\operatorname{sign}(q, \mathcal{L}X) = \operatorname{sign}\left(X, \bigotimes_{n=1}^{\infty} S_{q^n} T_{\mathbb{C}} \otimes \bigotimes_{n=1}^{\infty} \Lambda_{q^n} T_{\mathbb{C}}\right).$$

Therefore $\operatorname{sign}(q, \mathcal{L}X)$ *is a power series in* q *with integral coefficients and constant term (coefficient of* q^0 *)* $\operatorname{sign}(X)$.

Proof: It must be shown that

$$
\mathrm{ch}\Big(\bigotimes_{n=1}^{\infty} S_{q^n} T_{\mathbb{C}} \otimes \bigotimes_{n=1}^{\infty} \Lambda_{q^n} T_{\mathbb{C}}\Big) = \prod_{i=1}^{2k} \prod_{n=1}^{\infty} \left(\frac{(1 + q^n e^{-x_i})(1 + q^n e^{x_i})}{(1 - q^n e^{-x_i})(1 - q^n e^{x_i})} \right).
$$

However this is clear from the definitions of the bundles and the formulas for the Chern characters given in section 1.5.

Remark: It is well-known that a product $\prod (1 + u_i)$ converges absolutely provided that the series $\sum |u_i|$ converges, hence in our case only for $|q| < 1$. We have therefore only defined an equivariant signature formally, but it is meaningful as a power series.

We now consider the connection with elliptic genera: The constant term in the (not yet evaluated) expression for $\mathrm{sign}\,(q, \mathcal{L}X)$ is

$$
\left(2 \cdot \prod_{n=1}^{\infty} \frac{(1 + q^n)^2}{(1 - q^n)^2} \right)^{2k};
$$

we have therefore not formulated the expression stably. But $\mathrm{sign}\,(X, \mathcal{L}X)$ is, up to this normalization factor, the genus associated to

$$
\frac{x}{f(x)} = \frac{1}{2} \frac{x}{\tanh\,(x/2)} \cdot \prod_{n=1}^{\infty} \left(\frac{(1 + q^n e^x)(1 + q^n e^{-x})}{(1 - q^n e^x)(1 - q^n e^{-x})} : \frac{(1 + q^n)^2}{(1 - q^n)^2} \right).
$$

As discussed in Appendix I, Theorem 5.6, this is precisely the expansion of $x\sqrt{\wp(x) - e_1}$, where $q = e^{2\pi i \tau}$ and $\wp$ is the Weierstraß $\wp$-function for the lattice $L = 2\pi i(\mathbb{Z}\tau + \mathbb{Z})$ as described in section 2.1. We therefore immediately deduce the

Corollary: *We have*

$$
\mathrm{sign}\,(q, \mathcal{L}X) = \varphi(X) \cdot 2^{2k} \left(\prod_{n=1}^{\infty} \frac{(1 + q^n)^2}{(1 - q^n)^2} \right)^{2k}.
$$

Here $\varphi(X)$ is the elliptic genus of X, hence a modular form of weight $2k$ on $\Gamma_0(2)$ as described in Appendix I. We now investigate whether the correction factor is also a modular form. For the following, cf. Appendix I, §4: The Dedekind η-function is defined as

$$
\eta(\tau) = q^{1/24} \cdot \prod_{n=1}^{\infty} (1 - q^n)
$$

and is holomorphic in the upper half-plane $\mathfrak{h}$. Further, $\eta^{24} = \Delta$ is an element of weight 12 in $M_*^{\mathrm{cusp}}(\Gamma)$, the ring of cusp forms on the modular group $\Gamma = \mathrm{SL}_2(\mathbb{Z})$. Hence η

is a "modular form" of weight $1/2$ (the quotation marks being due to a problem with 24-th roots of unity).

Lemma: *The infinite product*

$$\left(2 \prod_{n=1}^{\infty} \frac{(1+q^n)^2}{(1-q^n)^2} \right)^{-4}$$

is a modular form of weight 4 *on* $\Gamma_0(2)$, *and is indeed precisely our modular form* ε.

Proof: We have

$$\prod_{n=1}^{\infty} \frac{(1-q^n)^2}{(1+q^n)^2} = \prod_{n=1}^{\infty} \frac{(1-q^n)^4}{(1-q^{2n})^2} = \frac{\eta(\tau)^4}{\eta(2\tau)^2}.$$

The rest follows from Appendix I, Corollary 4.11.

The reciprocal of the correction factor has the same weight $2k$ as the modular form $\varphi(X)$, but is only a modular form when $2k \equiv 0$ (4). We therefore have the

Theorem: *The equivariant signature is* $\operatorname{sign}(q, \mathcal{L}X) = \varphi(X) \cdot \varepsilon^{-k/2}$, *and is a modular function on* $\Gamma_0(2)$ *for* $\dim X = 4k \equiv 0$ (8).

Remark: By considering $X \times X$ if $\dim X \equiv 4$ (8) it follows that in every case $\operatorname{sign}(q, \mathcal{L}X)^2$ is a modular function on $\Gamma_0(2)$.

Example: Since $\varphi(P_2(\mathbb{C})) = \delta$, it follows that $\operatorname{sign}(q, \mathcal{L}P_2(\mathbb{C}))^2 = \delta^2/\varepsilon$.

Corollary: *For the quaternionic projective spaces we have:*

$$\operatorname{sign}(q, \mathcal{L}P_k(\mathbb{H})) = \begin{cases} 0, & \text{for } k \equiv 1 \ (2), \\ 1, & \text{for } k \equiv 0 \ (2). \end{cases}$$

For the Milnor manifolds H_{ij} *we have (taking* $j \geq i$ *as we may):*

$$\operatorname{sign}(q, \mathcal{L}H_{ij}) = \begin{cases} 0, & \text{for } i \equiv 1 \ (2) \text{ or } i \equiv j \ (2), \\ 1, & \text{for } i \equiv 0 \ (2) \text{ and } j = i+1. \end{cases}$$

In particular, for these manifolds the considered twisted signatures vanish (apart from $\operatorname{sign}(1, X) = \operatorname{sign}(X)$ *).*

Proof: Compare sections 1.7 and 3.2.

6.2 The expansion at the cusp 0

Let a lattice $L \subset \mathbb{C}$ be given and let $h(x) = x^2 + \cdots$ be an elliptic function with divisor $2 \cdot (0) - 2 \cdot (P)$ (then P must automatically be a two-division point of L);

then $f(x) = \sqrt{h(x)}$ with the normalization $f(x) = x + \cdots$ defines an unbranched two-sheeted covering $\mathbb{C}/\widetilde{L} \to \mathbb{C}/L$. Here $\widetilde{L} \subset L$ is a sublattice of index two with respect to which f is elliptic. Since $P \equiv -P$ (L) we have $\operatorname{div}(h(x)) = \operatorname{div}(h(-x))$, so $h(x)$ and $h(-x)$ coincide up to a constant factor. Because $h(x) = x^2 + \cdots$, this factor is 1 and h is an even function. Since $f(x) = \sqrt{h(x)} = x + \cdots$, f is odd and therefore $Q(x) = x/f(x) = 1 + \cdots$ is an even power series, which we can take as the characteristic power series for a genus.

This power series depends on the lattice (L, P) with distinguished two-division point. If f belongs to (L, P), then $f(\lambda x)/\lambda$ belongs to the lattice $(\frac{1}{\lambda} L, \frac{1}{\lambda} P)$. Hence the coefficients a_k of $Q(x)$ in the power series expansion $Q(x) = \sum_{k=o}^{\infty} a_k x^k$ are lattice functions of weight $-k$ for marked lattices. We shall now consider only lattices of the form $2\pi i(\mathbb{Z}\tau + \mathbb{Z})$. The passage from lattices $\mathbb{Z}\omega_2 + \mathbb{Z}\omega_1$ to these special lattices corresponds exactly to the passage from homogeneous lattice functions to modular forms. We thereby lose (up to a stretching factor) no information, since $\mathbb{Z}\omega_2 + \mathbb{Z}\omega_1 = \frac{\omega_1}{2\pi i} \cdot 2\pi i(\mathbb{Z}\frac{\omega_2}{\omega_1} + \mathbb{Z})$. In addition we suppose that τ lies in $\mathfrak{h} = \{z \in \mathbb{C} \mid \operatorname{Im}(z) > 0\}$, which we can always attain by permuting ω_2 and ω_1.

In our construction of the elliptic genus, for a lattice $L = 2\pi i(\mathbb{Z}\tau + \mathbb{Z})$ we take the point $P = \pi i$ and $h(x) = (\wp(x) - e_1)^{-1}$; the function $f(x) = (\wp(x) - e_1)^{-1/2}$ is elliptic for the lattice $2\pi i(\mathbb{Z} \cdot 2\tau + \mathbb{Z})$. We then obtain the Fourier expansion (cf. Appendix I, §5):

$$Q(x) = \frac{x}{f(x)} = \frac{1}{2} \frac{x}{\tanh(x/2)} \prod_{n=1}^{\infty} \frac{(1 + q^n e^x)(1 + q^n e^{-x})/(1 + q^n)^2}{(1 - q^n e^x)(1 - q^n e^{-x})/(1 - q^n)^2}.$$

For this choice of two-division point the passage from lattice functions to modular forms yields precisely modular forms on the congruence subgroup

$$\Gamma_0(2) = \{ \left(\begin{smallmatrix} a & b \\ c & d \end{smallmatrix} \right) \in \mathrm{SL}_2(\mathbb{Z}) \mid c \equiv 0 \,(2) \}$$

(cf. Appendix I, §4 and notice that $\Gamma_1(2) = \Gamma_0(2)$). Applied to a pair of basis vectors (ω_2, ω_1) of a lattice, for such a matrix we have

$$\begin{pmatrix} a & b \\ c & d \end{pmatrix} \begin{pmatrix} \omega_2 \\ \omega_1 \end{pmatrix} = \begin{pmatrix} a\omega_2 + b\omega_1 \\ c\omega_2 + d\omega_1 \end{pmatrix}$$

and the two-division point $\omega_1/2$ goes to the point $(c\omega_2 + d\omega_1)/2$, which is always equivalent to $\omega_1/2$ modulo L (since c is even and d must be odd due to the determinant condition).

If we take $f(x) = (\wp(x) - e_2)^{-1/2}$ in the construction of our elliptic genus, then we obtain (the product runs over all $m \geq 0$):

$$\frac{x}{f(x)} = \frac{x/2}{\sinh(x/2)} \frac{\displaystyle\prod_{n=2m+1} \left(1 - q^{n/2} e^x\right)\left(1 - q^{n/2} e^{-x}\right)/\left(1 - q^{n/2}\right)^2}{\displaystyle\prod_{n=2m+2} \left(1 - q^{n/2} e^x\right)\left(1 - q^{n/2} e^{-x}\right)/\left(1 - q^{n/2}\right)^2}.$$

The shift of indices under the translations $x \mapsto x + 2\pi i\tau$ and $x \mapsto x + 2\pi i$ (for which the sign changes) shows, that this function f is elliptic for the lattice $2\pi i(\mathbb{Z}\tau + \mathbb{Z} \cdot 2)$. Furthermore an analysis of the zeroes and poles shows that it has the demanded divisor. In addition the normalization is correct. The congruence subgroup $\Gamma_0(2)$ has two cusps, which can be represented by $\infty = \binom{1}{0}$ and $0 = \binom{0}{1}$. Modular forms F on this group must have a Fourier expansion without negative exponents at each of these cusps. At the cusp 0, one obtains this expansion by means of a matrix $A \in \mathrm{SL}_2(\mathbb{Z})$ with $A\binom{1}{0} = \binom{0}{1}$ (say $A = \left(\begin{smallmatrix} 0 & 1 \\ -1 & 0 \end{smallmatrix}\right)$), through the Fourier expansion of $F|_k A$ at the cusp ∞. This new modular form is now invariant under the action of

$$A^{-1}\Gamma_0(2)A = \Gamma^0(2) = \{\left(\begin{smallmatrix} a & b \\ c & d \end{smallmatrix}\right) \in \mathrm{SL}_2(\mathbb{Z}) \mid b \equiv 0\ (2)\}.$$

Notice that $\Gamma_0(2)$ and $\Gamma^0(2)$ are not normal subgroups of $\Gamma(1) = \mathrm{SL}_2(\mathbb{Z})$, since they are conjugate and do not coincide. Both have index 3 in $\Gamma(1)$. The matrices $S = \left(\begin{smallmatrix} 0 & -1 \\ 1 & 0 \end{smallmatrix}\right)$ ($\tau \mapsto -1/\tau$) and $T = \left(\begin{smallmatrix} 1 & 1 \\ 0 & 1 \end{smallmatrix}\right)$ ($\tau \mapsto \tau + 1$) generate $\Gamma(1)$. In the following figure, fundamental domains for $\Gamma_0(2)$ (parts labelled Id, S, ST) and $\Gamma^0(2)$ (parts labelled Id, S, T) are indicated, which we shall take as standard fundamental domains. Both have two cusps, 0 and ∞, which are exchanged by $\tau \mapsto -1/\tau$.

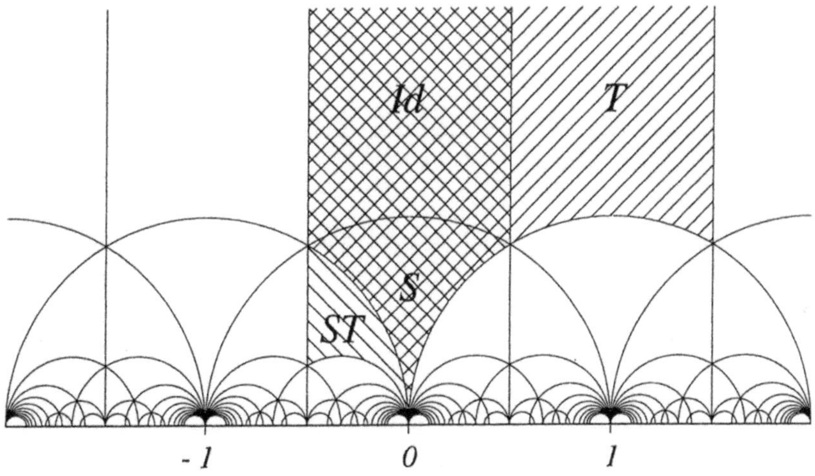

Some fundamental domains on the upper half-plane

Now our exchange of two-division points corresponds exactly to an exchange of the cusps (see Appendix I, Theorem 6.4). Only powers of $q^{1/2}$ enter into our expansion of $f(x)$, since $\left(\begin{smallmatrix} 1 & 2 \\ 0 & 1 \end{smallmatrix}\right) \in A^{-1}\Gamma_0(2)A = \Gamma^0(2)$, so $f|_k A$ is invariant under $\tau \mapsto \tau + 2$. The coefficients a_k of the Taylor expansion of $x/f(x)$ are therefore modular forms of weight k on $\Gamma^0(2)$ and have a Fourier expansion in $q^{1/2}$ at the cusp ∞. Since we shall later compare the elliptic genera for different two-division points with one another, we attempt to return to $\Gamma_0(2)$ by a further conjugation, which leaves the Fourier expansion essentially unchanged. The groups $\Gamma^0(2)$ and $\Gamma_0(2)$ are carried into one another by conjugation with the matrix $\begin{pmatrix} 2 & 0 \\ 0 & 1 \end{pmatrix}$:

$$\begin{pmatrix} \tfrac{1}{2} & 0 \\ 0 & 1 \end{pmatrix} \begin{pmatrix} a & b \\ c & d \end{pmatrix} \begin{pmatrix} 2 & 0 \\ 0 & 1 \end{pmatrix} = \begin{pmatrix} a & \tfrac{b}{2} \\ 2c & d \end{pmatrix} \in \Gamma_0(2)$$

if b is even. This corresponds to the passage from τ to 2τ, hence from q to q^2. Our expression above therefore becomes

$$\frac{x/2}{\sinh(x/2)} \prod_{n=1}^{\infty} \left(\frac{(1 - q^n e^x)(1 - q^n e^{-x})}{(1 - q^n)^2} \right)^{(-1)^{n-1}},$$

which is elliptic in x (after division by x) for the lattice $4\pi i(\mathbb{Z}\tau + \mathbb{Z})$ and has coefficients which are modular forms on $\Gamma_0(2)$.

Exercise: What happens for the third two-division point, i.e.

$$f(x) = \frac{1}{\sqrt{\wp(x) - e_3}} \ ?$$

For $f(x) = (\wp(x) - e_2)^{-1/2}$ we have obtained (product over $m \geq 0$)

$$\frac{x}{f(x)} = \frac{x/2}{\sinh(x/2)} \frac{\prod\limits_{n=2m+1} \left(1 - q^{n/2} e^x\right)\left(1 - q^{n/2} e^{-x}\right)/\left(1 - q^{n/2}\right)^2}{\prod\limits_{n=2m+2} \left(1 - q^{n/2} e^x\right)\left(1 - q^{n/2} e^{-x}\right)/\left(1 - q^{n/2}\right)^2}.$$

The corresponding genus for a manifold X^{4k} is then a modular form of weight $2k$ on $\Gamma^0(2)$, whose Fourier expansion is that of the usual genus φ for $f(x) = (\wp(x) - e_1)^{-1/2}$ at the cusp 0. If we replace τ by 2τ, we obtain a genus $\widetilde{\varphi}$ which again yields modular forms on $\Gamma_0(2)$:

$$\widetilde{\varphi}(X) = \left(\prod_{i=1}^{2k} \frac{x_i/2}{\sinh(x_i/2)} \prod_{n=1}^{\infty} \left(\frac{(1 - q^n e^{x_i})(1 - q^n e^{-x_i})}{(1 - q^n)^2} \right)^{(-1)^{n-1}} \right) [X].$$

We consider the normalization factor (product over $m \geq 0$):

$$\frac{\prod\limits_{n=2m+2} (1 - q^n)^2}{\prod\limits_{n=2m+1} (1 - q^n)^2} = \frac{\prod\limits_{n=2m+2} (1 - q^n)^4}{\prod\limits_{n=1}^{\infty} (1 - q^n)^2} = \prod_{n=1}^{\infty} \frac{(1 - q^{2n})^4}{(1 - q^n)^2}$$

$$= q^{-1/4} \frac{\eta(2\tau)^4}{\eta(\tau)^2}.$$

Here η is $q^{1/24} \cdot \prod_{n=1}^{\infty} (1 - q^n)$ and $\eta^{24} = q \cdot \prod_{n=1}^{\infty} (1 - q^n)^{24} = \Delta$ is again a cusp form of weight 12 on the full modular group. For a manifold of dimension $4k$ the normalization factor is

$$q^{-k/2} \left(\frac{\eta(2\tau)^4}{\eta(\tau)^2} \right)^{2k} ;$$

the expression inside the parentheses is a modular form of weight one, up to an indeterminacy in fourth roots of unity, since we have (cf. Appendix I, Corollary 4.11)

$$\left(\frac{\eta(2\tau)^4}{\eta(\tau)^2} \right)^4 = \tilde{\varepsilon} = \frac{\delta^2 - \varepsilon}{4} \in M_4(\Gamma_0(2)).$$

For k even, if one divides our expression for the genus by the modular form $\tilde{\varepsilon}^{k/2}$, the resulting quotient of two modular forms of weight $2k$ is the modular function

$$q^{-k/2} \cdot \left(\prod_{i=1}^{2k} \frac{x_i/2}{\sinh(x_i/2)} \prod_{n=1}^{\infty} \left((1 - q^n e^{x_i})(1 - q^n e^{-x_i}) \right)^{(-1)^{n-1}} \right) [X].$$

We now return to the $\hat{A}$-genus:

$$\hat{A}(X) = \left(\prod_{i=1}^{2k} \frac{x_i/2}{\sinh(x_i/2)} \right) [X].$$

Definition: *Let X be a differentiable manifold of dimension $2k$ and W a complex vector bundle over X. Then the **twisted $\hat{A}$-genus** $\hat{A}(X, W)$ is defined as*

$$\hat{A}(X, W) := \left(\mathrm{ch}\,(W) \cdot \hat{A}(TX) \right) [X]$$

$$= \left(\mathrm{ch}\,(W) \cdot \prod_{i=1}^{k} \frac{x_i/2}{\sinh(x_i/2)} \right) [X].$$

Remark: For a spin manifold X, $\hat{A}(X, W)$ is the index of a twisted Dirac operator and so is integral (cf. Appendix II).

Example: Let $X = S^{2k}$ be a sphere (so spin) and W a vector bundle over S^{2k}. Then

$$\hat{A}(S^{2k}, W) = \left(\hat{A}(TS^{2k}) \cdot \mathrm{ch}\,(W) \right) [S^{2k}] = \mathrm{ch}\,(W)[S^{2k}],$$

since $\hat{A}(TS^{2k}) = 1$ (spheres have stably trivial tangent bundles). As in section 6.1 it now follows that

$$\hat{A}(S^{2k}, W) = \frac{(-1)^{k+1}}{(k-1)!} c_k(W)[S^{2k}] \in \mathbb{Z},$$

so that we indeed obtain

$$c_k(W) \equiv 0 \ ((k-1)!).$$

In our expression above (product over $m \geq 0$)

$$q^{-k/2}\left(\prod_{i=1}^{2k} \frac{x_i/2}{\sinh(x_i/2)} \frac{\prod_{n=2m+1}(1-q^n e^{x_i})(1-q^n e^{-x_i})}{\prod_{n=2m+2}(1-q^n e^{x_i})(1-q^n e^{-x_i})}\right)[X],$$

we can recognize with a practiced eye that we are dealing with a twisted $\hat{A}$-genus, namely with

$$\frac{\widetilde{\varphi}(X)}{\widetilde{\varepsilon}^{k/2}} = q^{-k/2} \cdot \hat{A}\Big(X, \bigotimes_{n=2m+1} \Lambda_{-q^n} T_{\mathbb{C}} \otimes \bigotimes_{n=2m+2} S_{q^n} T_{\mathbb{C}}\Big).$$

By Witten, this can be interpreted as the equivariant index of the Dirac operator on loop space, twisted by a bundle that has no finite dimensional analogue. However, its meaning for us is simply the expansion at the other cusp of the equivariant signature of the loop space

$$\frac{\varphi(X)}{\varepsilon^{k/2}} = \mathrm{sign}\Big(X, \bigotimes_{n=1}^{\infty} S_{q^n} T_{\mathbb{C}} \otimes \bigotimes_{n=1}^{\infty} \Lambda_{q^n} T_{\mathbb{C}}\Big) = \mathrm{sign}\,(q, \mathcal{L}X).$$

6.3 The Witten genus

Let $L \subset \mathbb{C}$ be a lattice, and let

$$\sigma_L(x) = x \cdot \prod_{\omega \in L \setminus \{0\}} \left(\Big(1 - \frac{x}{\omega}\Big) \cdot \exp\Big(\frac{x}{\omega} + \frac{x^2}{2\omega^2}\Big)\right)$$

for $x \in \mathbb{C}$ be the Weierstraß σ-function for the lattice L. The infinite product converges uniformly on compact sets, due to the exponential convergence factors. These convergence factors are determined by logarithmically differentiating $\left(1 - \frac{x}{\omega}\right)$:

$$\frac{\left(1 - \frac{x}{\omega}\right)'}{1 - \frac{x}{\omega}} = \frac{-\frac{1}{\omega}}{1 - \frac{x}{\omega}} \equiv -\frac{1}{\omega} \cdot \Big(1 + \frac{x}{\omega}\Big) \equiv -\Big(\frac{1}{\omega} + \frac{x}{\omega^2}\Big) \quad (x^2).$$

The function $\sigma_L(x)$ is odd, so that the power series $Q(x) = x/\sigma_L(x) = 1 + a_2 x^2 + a_4 x^4 + \cdots$ is even. For this we have

$$\frac{\lambda x}{\sigma_L(\lambda x)} = \frac{x}{\sigma_{L/\lambda}(x)},$$

i.e. the coefficients a_r of x^r are multiplied by λ^r if the lattice is divided by λ. They are therefore homogeneous lattice functions of weight $-r$, and so modular forms with respect to $L = 2\pi i(\mathbb{Z}\tau + \mathbb{Z})$ of weight r on the full modular group $\mathrm{SL}_2(\mathbb{Z})$ (one must still argue that the a_r are holomorphic and determine the form of their Fourier expansions at the cusp ∞). Since there are no modular forms of weight 2, we have $a_2 = 0$.

This power series $Q(x)$ has a beautiful logarithm. Before giving this, we recall some important modular forms (cf. Appendix I, §3 and section 2.1; as usual, $q = e^{2\pi i \tau}$):

$$E_{2k}(\tau) = 1 + \frac{4k}{-B_{2k}} \cdot \sum_{n=1}^{\infty} \sigma_{2k-1}(n) \cdot q^n,$$

$$G_{2k}(\tau) = \frac{-B_{2k}}{4k} \cdot E_{2k} = \frac{-B_{2k}}{4k} + \sum_{n=1}^{\infty} \sigma_{2k-1}(n) \cdot q^n$$

$$= \frac{(2k-1)!}{2} \cdot \sum_{\omega \in L'} \frac{1}{\omega^{2k}} = \frac{(2k-1)!}{2} \cdot s_{2k}.$$

Here $L' = L \setminus \{0\}$. For $k > 1$, these are modular forms on the full modular group; G_2 (and so of course E_2) also converges,

$$G_2 = \frac{-B_2}{4} + \sum_{n=1}^{\infty} \sigma_1(n) \cdot q^n = -\frac{1}{24} + q + 3q^2 + \cdots.$$

We now have (cf. Appendix I, §5):

$$Q(x) = \exp\left(\sum_{k=2}^{\infty} \frac{2}{(2k)!} G_{2k}(\tau) \cdot x^{2k}\right).$$

As we have seen in section 1.8, the genera of complex manifolds for which all Chern classes apart from the highest vanish can be given precisely in terms of the logarithm of the characteristic power series. In this way, the so-called Witten genus φ_W associated to the characteristic power series $Q(x) = x/\sigma_L(x)$ yields Eisenstein series for such manifolds. The notation φ_W was proposed by Landweber, since the "W" reminds one of Weierstraß and/or Witten. For a compact, oriented, differentiable manifold X^{4k} we therefore have:

$$\varphi_W(X) = \left(\prod_{i=1}^{2k} \frac{x_i}{\sigma_L(x_i)}\right)[X]$$

is a modular form of weight $2k$.

Proposition: *The power series $Q(x)$ has the following product representation (cf. [Za86]):*

$$Q(x) = \frac{x/2}{\sinh(x/2)} \cdot \prod_{n=1}^{\infty} \frac{(1 - q^n)^2}{(1 - q^n e^x) \cdot (1 - q^n e^{-x})} e^{-G_2(\tau) \cdot x^2}.$$

Idea of the proof: Consider the zeroes and poles of both sides, for fixed τ. The σ-function has simple zeroes at all lattice points, and no poles. The function $\sinh(x/2)$ has zeroes at all multiples of $2\pi i$; the denominator vanishes if x modulo $2\pi i\mathbb{Z}$ is a proper multiple of $2\pi i\tau$. Together, on both sides we have simple poles at all non-zero lattice points. The functions are of course not elliptic, but rather are theta functions with respect to the lattice L. Therefore, they coincide a priori only up to a trivial theta function. The correct factor to yield equality is $e^{-G_2(\tau)\cdot x^2}$ (cf. Appendix I, §5).   ▱

In the product representation of $Q(x)$ some well-known terms appear, but the factor $e^{-G_2(\tau)\cdot x^2}$ is troublesome. For this reason, in Appendix I, §5, we consider a function Φ without this factor. Here we shall instead make the assumption that the first Pontrjagin class $p_1(X)$ of our manifold X vanishes in $H^4(X;\mathbb{Q})$, so if $p(X) = \prod_{i=1}^{2k}\left(1 + x_i^2\right)$ then $p_1(X) = \sum_{i=1}^{2k} x_i^2 = 0$, up to torsion. Then we obtain for the Witten genus:

$$\varphi_{\mathrm{W}}(X) = \left(\prod_{i=1}^{2k}\frac{x_i}{\sigma_L(x_i)}\right)[X]$$

$$= \left(\prod_{i=1}^{2k}\left(\frac{x_i/2}{\sinh(x_i/2)}\cdot\prod_{n=1}^{\infty}\frac{(1-q^n)^2}{(1-q^n e^{x_i})(1-q^n e^{-x_i})}e^{-G_2(\tau)\cdot x_i^2}\right)\right)[X]$$

$$= \left(\prod_{i=1}^{2k}\left(\frac{x_i/2}{\sinh(x_i/2)}\cdot\prod_{n=1}^{\infty}\frac{(1-q^n)^2}{(1-q^n e^{x_i})(1-q^n e^{-x_i})}\right)e^{-G_2(\tau)\cdot\sum_{i=1}^{2k} x_i^2}\right)[X]$$

$$= \left(\prod_{i=1}^{2k}\left(\frac{x_i/2}{\sinh(x_i/2)}\cdot\prod_{n=1}^{\infty}\frac{(1-q^n)^2}{(1-q^n e^{x_i})(1-q^n e^{-x_i})}\right)\right)[X].$$

As in the last section, we now consider the $\hat{A}$-genus with coefficients in a vector bundle E:

$$\hat{A}(X,E) = \left(\left(\prod_{i=1}^{2k}\frac{x_i/2}{\sinh(x_i/2)}\right)\cdot\mathrm{ch}\,(E)\right)[X].$$

We recognize the denominator in the product expansion of the Witten genus from the section on representations. In terms of the previous notation $S_t E = \sum_{r=0}^{\infty} S^r E\cdot t^r$ we have (only for $p_1(X) = 0$!):

$$\varphi_{\mathrm{W}}(X) = \hat{A}\left(X,\bigotimes_{n=1}^{\infty} S_{q^n} T_{\mathbb{C}}\right)\cdot\prod_{n=1}^{\infty}(1-q^n)^{4k}.$$

As always, $T_{\mathbb{C}}$ denotes the bundle $TX\otimes\mathbb{C}$. We shall rewrite the product on the right side. For this we again use the famous cusp form

$$\Delta = q\cdot\prod_{n=1}^{\infty}(1-q^n)^{24}$$

of weight 12, whose coefficients are the Ramanujan numbers. We already know two modular forms of weight 12, namely E_4^3 and E_6^2. Both begin with constant term 1 and are integral (since the Bernoulli numbers B_4 and B_6 have a 1 in the numerator); hence $E_4^3 - E_6^2$ is a cusp form, and we have after normalization:

$$\Delta = \frac{1}{1728} \cdot \left(E_4^3 - E_6^2 \right).$$

Here we have :

$$E_4 = 1 + 240 \cdot q + 2160 \cdot q^2 + \cdots,$$
$$E_6 = 1 - 504 \cdot q - 16632 \cdot q^2 + \cdots.$$

Despite the 1728, Δ has an integral Fourier expansion, as we see from the product expansion above. We substitute Δ into our expression for the Witten genus and so obtain:

$$\varphi_W(X) = q^{-\frac{4k}{24}} \cdot \hat{A}\left(X, \bigotimes_{n=1}^{\infty} S_{q^n} T_{\mathbb{C}}\right) \cdot \Delta^{\frac{4k}{24}}.$$

As we have already seen, φ_W always yields a modular form; $\Delta^{\frac{4k}{24}}$ is a modular form for $k \equiv 0 \ (6)$ and then

$$q^{-\frac{4k}{24}} \cdot \hat{A}\left(X, \bigotimes_{n=1}^{\infty} S_{q^n} T_{\mathbb{C}}\right)$$

becomes a modular function, otherwise we have difficulty with sixth roots of unity. The $\hat{A}$-genus is always integral up to powers of two in the denominator. For a spin manifold (this is equivalent to the vanishing of the second Stiefel-Whitney class) the $\hat{A}$-genus is the index of the Dirac operator and is therefore integral (cf. Appendix II). This also holds for the $\hat{A}$-genus with coefficients in suitable vector bundles (such as our $\bigotimes_{n=1}^{\infty} S_{q^n} T_{\mathbb{C}}$). Since the Fourier expansion of $q^{-\frac{4k}{24}} \cdot \Delta^{\frac{4k}{24}} = \prod_{n=1}^{\infty} (1 - q^n)^{4k}$ is also integral, we conclude the following

Proposition: *For a 4k-dimensional compact, oriented, differentiable manifold X with $p_1(X) = 0$ and $w_2(X) = 0$, the Witten genus φ_W yields a modular form of weight $2k$ with integral Fourier expansion.*

Example: Let X be as above, with $\dim X = 24$. The vector space of modular forms of weight 12 is spanned by E_4^3 and E_6^2, and so also by Δ and $E_4^3 - 744 \cdot \Delta$. The Witten genus $\varphi_W(X)$ of X is therefore a linear combination of these two modular forms:

$$\varphi_W(X) = q^{-1} \cdot \hat{A}\left(X, \bigotimes_{n=1}^{\infty} S_{q^n} T_{\mathbb{C}}\right) \cdot \Delta$$
$$= a \cdot \Delta + b \cdot \left(E_4^3 - 744 \cdot \Delta \right).$$

Why have we chosen precisely this basis?

There is the integral modular function

$$j(\tau) := \frac{E_4^3}{\Delta} = \frac{1}{q} + 744 + 196884 \cdot q + \cdots,$$

i.e. $j : \mathrm{SL}_2(\mathbb{Z})\backslash\mathfrak{h} \to P_1(\mathbb{C})$. If we identify $\mathrm{SL}_2(\mathbb{Z})\backslash\mathfrak{h}$ with $P_1(\mathbb{C})$, then j becomes a fractional linear transformation on $P_1(\mathbb{C})$, which is determined by its values at three points. For the j-function we may take these to be $j(i) = 1728$, $j\big(e^{2\pi i/3}\big) = 0$, and $j(\infty) = \infty$.

Dividing the above expression for $\varphi_{\mathrm{W}}(X)$ by Δ, we obtain:

$$a + b \cdot (j - 744) = \frac{b}{q} + a + \cdots = q^{-1} \cdot \hat{A}\Big(X, \bigotimes_{n=1}^{\infty} S_{q^n} T_{\mathbb{C}}\Big)$$

$$= q^{-1} \cdot \big(\hat{A}(X) + \hat{A}(X, T_{\mathbb{C}})q + \cdots\big)$$

$$\Rightarrow \quad a = \hat{A}(X, T_{\mathbb{C}}), \quad b = \hat{A}(X).$$

For $\hat{A}(X) \neq 0$ we now have the following formula:

$$\frac{q^{-1} \cdot \hat{A}\Big(X, \bigotimes_{n=1}^{\infty} S_{q^n} T_{\mathbb{C}}\Big)}{\hat{A}(X)} = \frac{\hat{A}(X, T_{\mathbb{C}})}{\hat{A}(X)} + j - 744.$$

The coefficients of $j - 744$ are very interesting. Here are the first four:

$$j - 744 = q^{-1} + 196884 \cdot q + 21493760 \cdot q^2 + 864299970 \cdot q^3 + \cdots.$$

These coefficients are in fact special sums of the dimensions of the irreducible representations of the Monster (see below). This is the largest sporadic finite simple group.

We now come to the

Prize Question: Does there exist a 24-dimensional, compact, orientable, differentiable manifold X with $p_1(X) = 0$, $w_2(X) = 0$, $\hat{A}(X) = 1$ and $\hat{A}(X, T_{\mathbb{C}}) = 0$?

Why are we interested in such a manifold? Under the above assumptions we have:

$$j - 744 = q^{-1} \cdot \hat{A}\Big(X, \bigotimes_{n=1}^{\infty} S_{q^n} T_{\mathbb{C}}\Big)$$

$$\equiv q^{-1} \cdot \hat{A}\big(X, (1 + qT_{\mathbb{C}} + q^2 S^2 T_{\mathbb{C}} + q^3 S^3 T_{\mathbb{C}})(1 + q^2 T_{\mathbb{C}})(1 + q^3 T_{\mathbb{C}})\big) \quad (q^3)$$

$$\equiv q^{-1}\big(\hat{A}(X) + \hat{A}(X, T_{\mathbb{C}})q + \hat{A}(X, S^2 T_{\mathbb{C}} + T_{\mathbb{C}})q^2$$

$$+ \hat{A}(X, S^3 T_{\mathbb{C}} + T_{\mathbb{C}} \otimes T_{\mathbb{C}} + T_{\mathbb{C}})q^3\big) \quad (q^3).$$

We therefore would have:

$$\hat{A}(X, S^2 T_{\mathbb{C}} + T_{\mathbb{C}}) = \hat{A}(X, S^2 T_{\mathbb{C}}) + \hat{A}(X, T_{\mathbb{C}})$$
$$= \hat{A}(X, S^2 T_{\mathbb{C}}) = 196884,$$
$$\hat{A}(X, S^3 T_{\mathbb{C}} + T_{\mathbb{C}} \otimes T_{\mathbb{C}} + T_{\mathbb{C}}) = \hat{A}(X, S^3 T_{\mathbb{C}}) + \hat{A}(X, T_{\mathbb{C}} \otimes T_{\mathbb{C}}) + \hat{A}(X, T_{\mathbb{C}})$$
$$= \hat{A}(X, S^3 T_{\mathbb{C}}) + \hat{A}(X, T_{\mathbb{C}} \otimes T_{\mathbb{C}})$$
$$= 21493760.$$

Now for each Riemannian metric on X there is a canonical nowhere vanishing section of $S^2 T_{\mathbb{C}}$. For a Riemannian metric on X yields a positive definite, symmetric bilinear form on each fibre of the tangent bundle of X, therefore a section of $S^2 T^*$. But T^* is isomorphic to T by means of this metric, and we obtain a section of $S^2 T$ and so also of $S^2 T_{\mathbb{C}}$. This section therefore defines a trivial line bundle E lying in $S^2 T_{\mathbb{C}}$, hence we can split $S^2 T_{\mathbb{C}}$ with respect to the metric:

$$S^2 T_{\mathbb{C}} = \widetilde{S}^2 T_{\mathbb{C}} \oplus E$$
$$\Rightarrow \quad \hat{A}(X, S^2 T_{\mathbb{C}}) = \hat{A}(X, \widetilde{S}^2 T_{\mathbb{C}}) + \hat{A}(X, E)$$
$$= \hat{A}(X, \widetilde{S}^2 T_{\mathbb{C}}) + \hat{A}(X)$$
$$\Rightarrow \quad \hat{A}(X, \widetilde{S}^2 T_{\mathbb{C}}) = 196883.$$

This number 196883 is the dimension of the smallest (non-trivial) representation of the Monster.

For the next coefficient 21493760 we have

$$T_{\mathbb{C}} \otimes T_{\mathbb{C}} = \Lambda^2 T_{\mathbb{C}} + S^2 T_{\mathbb{C}},$$
$$\Rightarrow \quad \hat{A}(X, S^3 T_{\mathbb{C}}) + \hat{A}(X, \Lambda^2 T_{\mathbb{C}}) = 21296876.$$

This is the dimension of the next largest irreducible representation of the Monster.

It would be marvelous to find a 24-dimensional manifold as in the prize question, on which in addition the Monster acts by diffeomorphisms. Such an action would lift to an action on the tangent bundle and its symmetric and exterior powers. In addition, the Monster would then also act on the cohomology of the twisted Dirac operators, whose indices are given by the twisted $\hat{A}$-genera. Hence one would have a relation between the Monster (more precisely, its irreducible representations), certain bundles on X and the coefficients of $j - 744$. Above all, one would possess a key to construct a great many representations of the Monster. Of course, the cohomology groups of the operators give only virtual representations so that an additive decomposition like in dimension 21296876 does not contradict the irreducibility.

Example: Following Landweber and Stong, we shall consider the Witten genus of complete intersections. Thus, let $V_1^{(d_1)}, \ldots, V_r^{(d_r)}$ be hypersurfaces of degree d_i in

$P_{n+r}(\mathbb{C})$. Assume the $V_i^{(d_i)}$ are smooth and intersect transversely (then their intersection is also smooth). The complete intersection $V_n = V_n^{(d_1,\dots,d_r)}$ then has complex dimension n. Denote by $x \in H^2(P_{n+r}(\mathbb{C}); \mathbb{Z})$ the canonical generator, the Poincaré dual of $P_{n+r-1}(\mathbb{C}) \subset P_{n+r}(\mathbb{C})$. Then the total Chern class of V_n is (cf. section 3.1):

$$c(V_n) = (1 + x)^{n+r+1} \cdot (1 + d_1 x)^{-1} \cdots (1 + d_r x)^{-1}|_{V_n} .$$

For the Witten genus of V_n we obtain the formula:

$$\varphi_{\mathrm{W}}(V_n) = \left(\frac{x^{n+r+1} \cdot \sigma(d_1 x) \cdots \sigma(d_r x)}{\sigma(x)^{n+r+1} \cdot d_1 x \cdots d_r x} \right)[V_n]$$

$$= \text{coefficient of } x^{n+r} \text{ in } \left(\frac{x^{n+r+1} \cdot \sigma(d_1 x) \cdots \sigma(d_n x)}{\sigma(x)^{n+r+1}} \right)$$

$$= \mathrm{res}_0\left(\frac{\sigma(d_1 x) \cdots \sigma(d_n x)}{\sigma(x)^{n+r+1}} \right).$$

We shall now determine this residue when $p_1(V_n) = 0$. We have :

$$p_1(V_n) = \left(n + r + 1 - \left(d_1^2 + \cdots + d_r^2 \right) \right) \cdot x^2|_{V_n} ,$$

as one easily concludes from the formula for the total Chern class. Hence we obtain:

$$p_1(V_n) = 0 \quad \text{is equivalent to} \quad n + r + 1 = d_1^2 + \cdots + d_r^2.$$

Now $\sigma(d_1 x) \cdots \sigma(d_r x)/\sigma(x)^{n+r+1}$ is an elliptic function if the number of zeroes and poles of the numerator is equal to that of the denominator, and the sum of the zeroes is equal to the sum of the poles (modulo the lattice). Now the σ-function has only simple zeroes at the lattice points, hence the factors of the numerator have zeroes at all d_i-division points, giving a total of $d_1^2 + \cdots + d_r^2$ simple zeroes. The denominator has an $(n + r + 1)$-fold zero at all lattice points. In case $p_1(V_n) = 0$, we have $n + r + 1 = d_1^2 + \cdots + d_r^2$; in addition, the sum of the zeroes is zero for each factor. Therefore the above quotient is an elliptic function, having a pole only at the origin (modulo the lattice). Hence its residue at the origin is equal to zero, since for elliptic functions the sum of the residues is always zero. Therefore:

$$p_1(V_n) = 0 \quad \text{implies} \quad \varphi_{\mathrm{W}}(V_n) = 0.$$

In particular, the prize question cannot be answered with complete intersections, since for $p_1(V_n) = 0$ the twisted $\hat{A}$-genera above vanish.

6.4 The Witten genus and the Lie group E_8

We shall assume that M is a $4k$-dimensional compact, orientable, differentiable manifold for which all composite Pontrjagin numbers vanish, i.e. with $p = 1 + p_1 + \cdots + p_k$ we have:

$$(p_{i_1} \cdots p_{i_r})[M] = 0 \qquad \text{for all } i_1 + \cdots + i_r = k \text{ with } r > 1.$$

The only remaining Pontrjagin number is $p_k[M]$. In addition, as earlier we shall require that $p_1 = 0$. In this situation, as expected, we obtain a much simplified formula for the Witten genus, since we can calculate modulo the Pontrjagin·classes $p_1, \ldots, p_{k-1}$ and so can simply apply the Cauchy lemma of section 1.8 for the calculation of the genus. We recall that the characteristic power series has a representation as

$$\frac{x}{\sigma_L(x)} = \exp\Big(\sum_{r=2}^{\infty} \frac{-B_{2r}}{2r \cdot (2r)!} E_{2r}(\tau) \cdot x^{2r}\Big);$$

hence $\varphi_W(M)$ is a constant multiple of E_{2k}. The constant term of E_{2k} is 1, while for $\varphi_W(M)$ it is $\hat{A}(M)$; therefore

$$\varphi_W(M) = \hat{A}(M) \cdot E_{2k}.$$

This is not very surprising, since under our assumptions on the Pontrjagin numbers every genus is a multiple of $\hat{A}$.

Example: For $k = 2$, so that M is an 8-dimensional manifold, we have :

$$\varphi_W(M) = \hat{A}(M) \cdot E_4$$
$$\Rightarrow \quad \frac{\varphi_W(M)}{\Delta^{1/3}} = \hat{A}(M) \cdot \frac{E_4}{\Delta^{1/3}} = \hat{A}(M) \cdot j^{1/3}$$
$$= \hat{A}(M) \cdot q^{1/3}\big(1 + 248 \cdot q + 4124 \cdot q^2 + \cdots\big).$$

The power series for the cube root of j (normalized as here) could have rational coefficients, having a three in the denominator, but the cube root of j is in fact an integral power series.

Suppose that we had an 8-dimensional manifold M with $p_1(M) = 0$ and $\hat{A}(M) = 1$. Then the above conditions are fulfilled, since the only Pontrjagin numbers in dimension 8 are p_1^2 and p_2. For such a manifold we have

$$\frac{\varphi_W(M)}{\Delta^{1/3}} = q^{-1/3}\big(1 + 248 \cdot q + 4124 \cdot q^2 + \cdots\big),$$

so by the previous formula (for $p_1 = 0$):

$$\frac{\varphi_W(M)}{\Delta^{1/3}} = q^{-1/3} \cdot \hat{A}\Big(M, \bigotimes_{n=1}^{\infty} S_{q^n} T_{\mathbb{C}}\Big)$$
$$\Rightarrow \quad \hat{A}(M, T_{\mathbb{C}}) = 248, \quad \hat{A}(M, T_{\mathbb{C}}) + \hat{A}(M, S^2 T_{\mathbb{C}}) = 4124,$$
$$\Rightarrow \quad \hat{A}(M, S^2 T_{\mathbb{C}}) = 3876 \quad \text{and} \quad \hat{A}(M, \widetilde{S}^2 T_{\mathbb{C}}) = 3875.$$

Here one could again conjecture a relation to the irreducible representations of a group. For this we shall consider the exceptional Lie group E_8 (despite a close connection, the identical notation with the Eisenstein series E_8 is accidental). It has dimension 248 and therefore possesses a representation in this dimension, namely the adjoint representation on its Lie algebra. At the same time, this is the smallest irreducible representation of E_8. The next dimension in which an irreducible representation exists is precisely 3875. However, one cannot hope that E_8 acts non-trivially on our manifold, since then an $S^1 \subset E_8$ would also act non-trivially (each element lies in a maximal torus, hence also in an S^1). But this contradicts $\hat{A}(M) = 1$, since we have (cf. [AtHi70]) the

Theorem (Atiyah-Hirzebruch): *If the circle S^1 acts non-trivially on a connected spin manifold, then $\hat{A}(M) = 0$.*   ⌷

But one could still hope that perhaps a large discrete subgroup of E_8 acts on a manifold of the type assumed to exist in this example. As already noted above, under our assumptions on the Pontrjagin numbers of the manifold, each genus is a multiple of the $\hat{A}$-genus; in particular, we have (by the Cauchy lemma):

$$\operatorname{sign}(M) = -2^{2k+1}\left(2^{2k-1} - 1\right) \cdot \hat{A}(M).$$

Since the $\hat{A}$-genus is always integral up to a power of two in the denominator, $2^{2k-1} - 1$ must divide the signature of M. If in addition M is a spin manifold, so that the second Stiefel-Whitney class $w_2(M)$ is equal to zero, then the twisted $\hat{A}$-genera are integral; since this is likewise true of $\Delta^{k/6}$, also

$$q^{-k/6} \cdot \hat{A}\Big(M, \bigotimes_{n=1}^{\infty} S_{q^n} T_{\mathbb{C}}\Big) \cdot \Delta^{k/6} = \varphi_{\mathrm{W}}(M) = \hat{A}(M) \cdot E_{2k}$$

is integral. We therefore obtain from

$$E_{2k}(\tau) = 1 + \frac{4k}{-B_{2k}} \cdot \sum_{n=1}^{\infty} \sigma_{2k-1}(n) \cdot q^n$$

for $n = 1$ the divisibility condition:

$$\hat{A}(M) \equiv 0 \quad \text{(numerator of the reduced representation of } B_{2k}/4k).$$

For $k = 2, 3$ this numerator is equal to one (E_4 and E_6 are integral modular forms), but for larger k this represents a proper divisibility. If we combine this with the above condition on the signature, then we have:

$$\operatorname{sign}(M) \equiv 0 \quad \left(2^{2k+1}\left(2^{2k-1} - 1\right) \cdot \operatorname{numerator}\left(\tfrac{B_{2k}}{4k}\right)\right).$$

In the derivation of this result we have proceeded very crudely. More precisely, we have only used the integrality of the coefficient of q in $\varphi_{\mathrm{W}}(M)$; hence it suffices to require $\hat{A}(M, T_{\mathbb{C}}) \in \mathbb{Z}$ instead of $w_2 = 0$ to deduce the divisibility statement.

We can obtain a small improvement of this result: For a spin manifold M, the $\hat{A}$-genus $\hat{A}(M)$ is always even if the dimension of M is divisible by 4 but not by 8 (cf. section 8.1). Putting $a_k = 1$ if k is even and $a_k = 2$ if k is odd, we therefore have:

$$\operatorname{sign}(M) \equiv 0 \quad \left(a_k 2^{2k+1}\left(2^{2k-1} - 1\right) \cdot \operatorname{numerator}\left(\tfrac{B_{2k}}{4k}\right)\right).$$

Example: For $k = 1$ this means that $\operatorname{sign}\left(M^4\right) \equiv 0$ (16). This result is sharp, since e.g. a $K3$-surface has signature 16.

Now that we have derived a lovely result under strong assumptions on the Pontrjagin numbers of the manifold, obviously the question remains whether such manifolds exists. Milnor and Kervaire have reported on this problem at the 1958 International Congress of Mathematicians (cf. [MiKe58]). They investigated a particular kind of manifold:

Definition: *A manifold is called **parallelizable** if its tangent bundle is trivial; it is called **almost-parallelizable** if its tangent bundle is trivial on the complement of a point.*

For an almost-parallelizable manifold M^{4k} we obviously have $w_2 = 0$, since w_2 can be defined via obstruction theory on the 3-skeleton of a triangulation of the manifold, and this triangulation can be chosen so that the tangent bundle of M is trivial on the $(4k - 1)$-skeleton, hence there is no obstruction to a section over this skeleton. The same argument also yields $p_i = 0$ for $i < k$. Moreover, such a manifold is automatically orientable, since also the obstruction w_1 to orientability is zero.

Since all our conditions on characteristic classes are fulfilled by almost-parallelizable manifolds, our divisibility theorem holds for such manifolds. The condition of being almost-parallelizable is in fact stronger than we require, but Milnor and Kervaire have constructed almost-parallelizable manifolds whose signature not only satisfy the above divisibility relation, but in fact yield an equality:

Theorem: *There are almost-parallelizable manifolds M_0^{4k} with*

$$\operatorname{sign}\left(M_0^{4k}\right) = a_k 2^{2k+1}\left(2^{2k-1} - 1\right) \cdot \operatorname{numerator}\left(\tfrac{B_{2k}}{4k}\right).$$

The divisibility statement is therefore sharp. For the Witten genus of M_0^{4k} this means that:

$$q^{-\frac{k}{6}} \cdot \hat{A}\left(M_0^{4k}, \bigotimes_{n=1}^{\infty} S_{q^n} T_{\mathbb{C}}\right) = \frac{\hat{A}\left(M_0^{4k}\right) \cdot E_{2k}}{\Delta^{k/6}}$$

$$= -a_k \cdot \operatorname{numerator}\left(\tfrac{B_{2k}}{4k}\right) \frac{E_{2k}}{\Delta^{k/6}}.$$

For M_0^8 we get $j^{1/3}$, hence for $M_0^8 \times M_0^8 \times M_0^8$ we obtain the modular function j; but this does not answer our prize question, since we want to attain instead $j - 744$.

Now to the construction of M_0^{4k}, by means of plumbing!

6.5 Plumbing of manifolds

The manifold M_0^{4k} of the last section will be constructed by judiciously gluing together several copies of the disk bundle associated to the tangent bundle of S^m. The gluing instructions are given by a suitable graph. This method of construction is called plumbing (cf. [MiKe58], [KeMi63], [Br72]).

We therefore consider $D(S^m) = \{x \in TS^m \mid |x| \le 1\}$. One is given a graph as plan for the construction, such as the diagram for E_8:

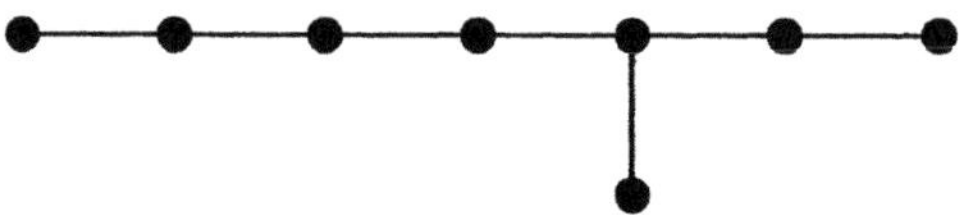

Diagram for E_8

For each node (or vertex) one takes a copy of $D(S^m)$, and joins together two of these copies if there is an edge in the graph between the corresponding nodes. The manner in which the copies of $D(S^m)$ are joined is shown here in the case $m = 1$:

Plumbing of disk bundles

The boundary of the new manifold in this example is obviously an S^1, and the manifold is orientable. In the general case plumbing also yields an orientable manifold X^{2m} with boundary. One can now easily determine the signature of this manifold. We recall that the signature of a $4k$-dimensional, oriented manifold X with boundary is defined as the signature of the quadratic form belonging to the intersection form on $H^{2k}(X, \partial X; \mathbb{Z}) \cong H_{2k}(X; \mathbb{Z})$, i.e.

$$a \in H^{2k}(X, \partial X; \mathbb{Z}) \mapsto (a \cdot a)[X, \partial X].$$

In $D(S^m)$ a canonical basis for the middle homology is given by the zero-section of the tangent bundle:

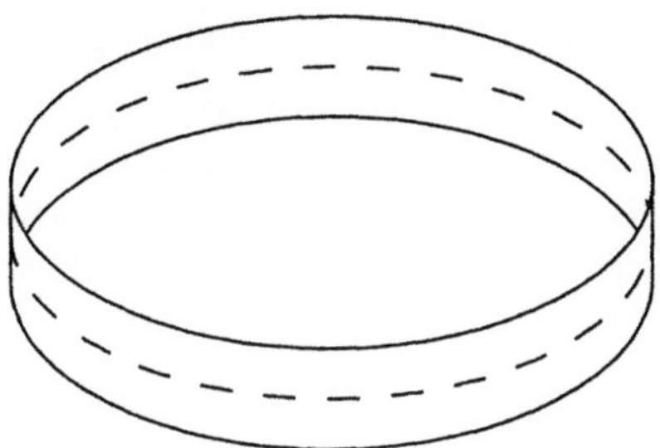

Zero-section of the disc bundle $D(S^1)$

Let $X_{E_8}^{2m}$ be the resulting manifold in the construction using E_8. The cycles $s_1,\ldots,s_8$ intersect one another exactly as indicated in the E_8-diagram ($i \neq j$):

$$s_i s_j = \begin{cases} 0, & \text{if no edge joins vertices } i \text{ and } j, \\ \pm 1, & \text{otherwise.} \end{cases}$$

The signs can be determined by the choice of an orientation of the s_i. The resulting matrix is symmetric for even m, anti-symmetric for odd m. We now consider the self-intersection numbers. For the S^1 the zero-section can be deformed so that it is nowhere zero, hence we obtain self-intersection number zero. This is not the case for S^2, since the deformation of the zero-section is given by a vector field on S^2, which is well-known to always have two zeroes (the Euler number) when counted correctly. We therefore have self-intersection number 2 here. In general, the Euler number of a sphere S^m is equal to 2 for even m and to 0 for odd m, hence:

$$s_i s_i = \begin{cases} 2, & \text{if } m \text{ is even}, \\ 0, & \text{otherwise.} \end{cases}$$

For the diagram E_8 we obtain the following matrix when $m = 2k$:

$$\begin{pmatrix} 2 & 1 & & & & & & \\ 1 & 2 & 1 & & & & & \\ & 1 & 2 & 1 & & & & \\ & & 1 & 2 & 1 & & & \\ & & & 1 & 2 & 1 & & 1 \\ & & & & 1 & 2 & 1 & \\ & & & & & 1 & 2 & \\ & & & & 1 & & & 2 \end{pmatrix}$$

It follows that the intersection form is positive definite with determinant 1.

In our example with S^1 we obtain a 2-dimensional manifold $X_{E_8}^2$ whose first (i.e. middle) homology is generated by eight cycles, i.e. $b_1 = 8$. If we collapse the boundary, an S^1, to a point, a Riemann surface of genus four (Euler number -6) results. In other words: $X_{E_8}^2$ is a Riemann surface of genus four with a hole.

In the general case, the boundary V^{2m-1} of X^{2m} is a homology sphere if the determinant of the intersection form of X is equal to ± 1. For $m > 2$ the boundary is also simply connected, hence for $m \neq 2$ is even a homotopy sphere (all homotopy groups of V^{2m-1} are the same as for the sphere S^{2m-1}). Now in 1960 Stephen Smale proved the following theorem (cf. [Sm60]):

Theorem (Smale): *Each homotopy sphere V^k with $k > 4$ is homeomorphic to the sphere S^k.*

For us this means that the boundary V^{2m-1} for $m \neq 2$ is therefore homeomorphic to the sphere S^{2m-1}. Now the disk bundle $D(S^m)$ is a parallelizable manifold: If $\pi : D(S^m) \to S^m$ is the projection we have $T(D(S^m)) = \pi^*(T(S^m) \oplus T(S^m))$ and $T(S^m) \oplus T(S^m)$ is a stably-trivial bundle over S^m with rank $2m$, so it is trivial. Hence the manifold $X_{E_8}^{2m}$ obtained from $D(S^m)$ by plumbing is again parallelizable.

We have thereby found an example of a so-called exotic sphere, i.e. V^{2m-1} is homeomorphic, but not diffeomorphic to S^{2m-1}! For if V^{2m-1} were diffeomorphic to S^{2m-1}, then

$$X_{E_8}^{2m} \cup_{S^{2m-1}} D^{2m}$$

would be a $2m$-dimensional manifold which is parallelizable in the complement of a point, hence almost-parallelizable. However the signature does not change when one attaches a disk, and so for $m = 2k$ it remains equal to 8. From our earlier considerations on the divisibility properties of the signature of almost-parallelizable manifolds we obtain the result:

$$a_k 2^{2k+1} \left(2^{2k-1}\right) \cdot \text{numerator}\left(\tfrac{B_{2k}}{4k}\right) \quad \text{divides} \quad 8.$$

But for $k = 2$ the left side is $8 \cdot 28$. Contradiction!

The Milnor spheres V^{2k-1} are therefore not diffeomorphic to S^{4k-1}, and

$$X_{E_8}^{2m} \cup_{S^{2m-1}} D^{2m}$$

only has a differentiable structure on the complement of a point. We now denote by bP_{4k} the set of diffeomorphism classes of oriented spheres Σ^{4k-1} which bound compact parallelizable manifolds (respecting orientations). This is a group with the connected sum as group operation. One forms the connected sum of two manifolds by joining the two manifolds with a handle. For $k \geq 2$ this group is generated by the V^{4k-1}. The order of the group is precisely

$$n_k = \frac{1}{8}\left(a_k 2^{2k+1}\left(2^{2k-1} - 1\right) \cdot \text{numerator}\left(\tfrac{B_{2k}}{4k}\right)\right),$$

hence the boundary connected sum of n_k copies of the manifold $X_{E_8}^{4k}$ has the standard sphere as boundary, to which one can glue a disk smoothly. Since the signature of two manifolds adds under the formation of their connected sum, the resulting manifold is therefore the desired manifold M_0^{4k}.

7 Riemann-Roch and elliptic genera in the complex case

7.1 Elliptic genera of level N

Let $\tau \in \mathfrak{h}$ and $L = 2\pi i(\mathbb{Z}\tau + \mathbb{Z})$ be a lattice. We seek a function $h(x)$ on $\mathbb{C}$, which is elliptic with respect to L and on $\mathbb{C}/L$ has a zero of order N at the origin and a pole of the same order at a point P. In order that such a function exist it is necessary that $N \cdot 0 - N \cdot P \equiv 0 \, (L)$, hence the point P must be a non-zero N-division point of $\mathbb{C}/L$. These N-division points are precisely the points P of the form:

$$P = 2\pi i \frac{k\tau + l}{N} \neq 0 \quad \text{for } k, l \in \mathbb{Z}.$$

If we normalize the function h by $h(x) = x^N + \cdots$, then it is uniquely determined for a fixed N-division point P (cf. Appendix I, §6). Under these assumptions, we can extract a unique N-th root f with the normalization

$$f(x) = \sqrt[N]{h(x)} = x + \cdots.$$

This is elliptic with respect to a sublattice $\widetilde{L}$ of index N in L. For fixed k, l the power series

$$Q(x) = \frac{x}{f(x)} = 1 + b_1 x + b_2 x^2 + \cdots$$

is therefore uniquely specified. The coefficients b_r depend on τ; they are modular forms of weight r for a certain modular group (cf. Appendix I, §6). This is not the full modular group $\mathrm{SL}_2(\mathbb{Z})$, since the chosen N-division point P must transform correctly. For $k = 0$ and $(l, N) = 1$ (P a primitive N-division point) one obtains as modular group for the b_r the group

$$\Gamma_1(N) = \left\{ \left(\begin{smallmatrix} a & b \\ c & d \end{smallmatrix} \right) \in \mathrm{SL}_2(\mathbb{Z}) \mid \left(\begin{smallmatrix} a & b \\ c & d \end{smallmatrix} \right) \equiv \left(\begin{smallmatrix} 1 & b \\ 0 & 1 \end{smallmatrix} \right) (N) \right\}.$$

We now consider the genus defined by this power series Q. Thus let X be a complex manifold of dimension d, with total Chern class given by:

$$c(X) = 1 + c_1(X) + \cdots + c_d(X) = (1 + x_1) \cdots (1 + x_d), \qquad c_i \in H^{2i}(X; \mathbb{Z}).$$

The genus $\varphi_{N,p}$ corresponding to Q is then

$$\varphi_{N,p}(X) = \left(\prod_{i=1}^{d} \frac{x_i}{f(x_i)} \right) [X].$$

This is an expression in the coefficients of $Q(x)$ of weight d, hence a modular form of weight d, on $\Gamma_1(N)$ if $k = 0$, i.e. if $P = 2\pi i l/N$ with $0 < l < N$. In fact, if $(l, N) = n \neq 1$ then we obtain the larger modular group $\Gamma_1(N/n) \supset \Gamma_1(N)$. For $k \neq 0$ different modular groups arise.

This genus is called an elliptic genus of level N. For $N = 2$ this is our original elliptic genus; the generalization to arbitrary N was already proposed by Witten.

In the case $N = 2$ one has the involution $x \mapsto -x$ on $\mathbb{C}/L$. This leaves the origin and P fixed and so also h, since we have normalized h to begin with x^2. Hence h is even, as well as also $Q(x) = x/\sqrt{h(x)}$. For the calculation of the corresponding genus one therefore needs only the elementary symmetric functions in the squares of the x_i, hence the Pontrjagin classes. Therefore one can also define the elliptic genus of level two for differentable manifolds. For $k = 0$ one obtains in this case modular forms on $\Gamma_1(2) = \Gamma_0(2)$.

7.2 The values at the cusps

For the elliptic genera of level two the values at the cusps of $\Gamma_0(2)$ are the well-known signature (L-genus) and $\hat{A}$-genus. We are therefore also interested in the values at the cusps of $\Gamma_1(N)$ for the elliptic genera of level N. One obtains these values by introducing the local uniformizer $q = e^{2\pi i \tau}$ for the cusp ∞, for example, and considering the Fourier expansion $\alpha_0 + \alpha_1 q + \cdots$ of the modular form at ∞. The value at ∞ is then α_0. For modular forms of odd weight this is only well-defined up to sign (cf. Appendix I, §1). For the values at the other cusps, one transforms them to ∞ and then proceeds analogously. We shall now investigate these values.

Thus let X be a compact, complex manifold of complex dimension d. Then the Hodge numbers $h^{p,q}$ are defined (cf. section 5.4), and as previously considered the p-th holomorphic Euler number is defined as:

$$\chi^p := \sum_{q=0}^{d} (-1)^q \cdot h^{p,q}.$$

χ^0 is just the arithmetic genus, and we have $\chi^p = (-1)^d \chi^{d-p}$. We have already shown that χ^p is the index of a suitable elliptic complex (the Dolbeault complex). We further defined

$$\chi_y(X) = \sum_{p=0}^{d} \chi^p(X) \cdot y^p;$$

the Atiyah-Singer index theorem yields

$$\chi_y(X) = \prod_{i=1}^{d} \left(x_i \frac{1 + y \cdot e^{-x_i}}{1 - e^{-x_i}} \right) [X].$$

Since the power series on the right side does not begin with 1, we must normalize it in order to obtain a genus:

$$\frac{\chi_y(X)}{(1+y)^d} = \prod_{i=1}^{d}\Big(x_i\,\frac{1+y\cdot e^{-x_i}}{1-e^{-x_i}}\cdot\frac{1}{1+y}\Big)[X].$$

Unfortunately we no longer obtain $\chi_{-1}(X)$, the Euler number $e(X)$ of X. For $y=1$ we have $\chi_1(X) = \text{sign}\,(X)$ which is already defined for differentiable manifolds.

We now consider a different genus. Let $K = \Lambda^d T^*$ be the canonical line bundle of X, and D an arbitrary divisor on X. We define

$$\chi(X,D) = \sum_{i=0}^{d} (-1)^i \dim H^i(X;\Omega^0(D)),$$

where $\Omega^0(D)$ is the sheaf of meromorphic functions f with $(f)+D \geq 0$ (resp. germs of holomorphic sections of the line bundle associated to D). The Riemann-Roch theorem yields:

$$\chi(X,D) = (\text{ch}\,(D)\cdot \text{td}\,(TX))[X]$$

$$= \Big(e^{c_1(D)}\cdot \prod_{i=1}^{d}\frac{x_i}{1-e^{-x_i}}\Big)[X].$$

Since $c_1(K) = -(x_1 + \cdots + x_d)$, we have:

$$\chi(X,K^r) = \Big(\prod_{i=1}^{d} e^{-rx_i}\frac{x_i}{1-e^{-x_i}}\Big)[X].$$

Hence $\chi(X,K^r)$ is the genus associated to the power series

$$Q(x) = e^{-rx}\cdot \frac{x}{1-e^{-x}}\,.$$

One can also consider $r \in \mathbb{Q}$ formally and so obtains a genus from the same formula, which is in general no longer integral. Sometimes K is the N-th power of a line bundle (which holds precisely when $c_1(X) \equiv 0\,(N)$). In this case, the genus $\chi(X,K^r)$ corresponding to $r = k/N$ is obviously integral. For $N = 2$ this means the purely topological condition $w_2(X) = 0$, but for $N > 2$ we need the complex structure.

We now turn to the values at the cusps. Let $P \neq 0$ be a fixed N-division point in $\mathbb{C}/L$, $L = 2\pi i(\mathbb{Z}\tau + \mathbb{Z})$. Without restriction we may assume that $P = 2\pi i/N$. For a different N-division point we obtain a different modular group. Just as is the case for level two, the choice of a different point P corresponds to the expansion at a different cusp (cf. Appendix I, §6). Given L and P, we have defined a power series $f(x) = \sqrt[N]{h(x)}$ in the above discussion. The elliptic genus of level N corresponding to $Q(x) = x/f(x)$ yields modular forms on $\Gamma_1(N)$ of weight d. For the values at the cusps, we can formulate the following assertion:

Theorem: *The values of the elliptic genus of level N at the cusps of $\Gamma_1(N)$ are:*

1) $\chi_y(X)/(1+y)^d$, *where* $-y = e^{2\pi i l/N} \neq 1$ *is an N-th root of unity with* $0 < l < N$.

2) $\chi(X, K^{k/N})$ *with* $0 < k < N$.

Proof: The expansion at some cusp of the elliptic genus for a fixed N-division point P corresponds to the expansion at the cusp ∞ for some other choice of N-division point. Therefore the set of values at the cusps for a fixed N-division point is the same as the set of values at the cusp ∞ for all N-division points $P = 2\pi i(k\tau + l)/N$.

These can be easily obtained from Appendix I, Theorem 6.4 (i). For $k = 0$ i.e. $P = 2\pi i l/N$ we get the value $\chi_y(X)/(1+y)^d$, where $-y = e^{2\pi i l/N}$, for $0 < k < N$ and l arbitrary we get $\chi(X, K^{k/N})$. □

Remark: The values at the cusps are in general not all distinct, since the number $\operatorname{cu}(N)$ of cusps of $\Gamma_1(N)$ is given, for $N = 3$ or $N > 4$, by (cf. Appendix I, Corollary 7.11)

$$\operatorname{cu}(N) = \frac{1}{2} \cdot \sum_{m|N} \varphi(m)\varphi(N/m),$$

where φ denotes the Euler φ-function. Hence for a prime $N = p \neq 2$ there are $p - 1$ cusps, but we have listed $2(p - 1)$ values. This comes about since these values are only well-defined up to sign. For $N = 2$, in case 1) only $y = 1$ occurs, i.e. the signature; in case 2) only $k = 1$, the $\hat{A}$-genus, is possible. These are the genera which we have already obtained at the two cusps of $\Gamma_1(2) = \Gamma_0(2)$.

7.3 The equivariant case and multiplicativity

There are several classical results about these values at the cusps. Assume that the circle S^1 acts non-trivially by holomorphic automorphisms on our complex manifold X. Then S^1 acts on the finite dimensional vector spaces $H^{p,q}(X)$. We thereby obtain a representation of S^1. The corresponding character is a finite Laurent series in $q = e^{2\pi i\tau} \in S^1$, $\tau \in \mathbb{R}$. Taken together, these representations yield as character a finite Laurent series $\chi_y(q, X)$ in q with $\chi_y(\mathrm{id}, X) = \chi_y(X)$. As we have seen in section 5.4, $\chi_y(q, X) \equiv \chi_y(X)$ is constant.

If $c_1(X) \equiv 0\ (N)$, i.e. $K = L^N$ for a line bundle, then $\chi(X, K^{k/N}) = \chi(X, L^k)$ is integral. If S^1 acts non-trivially on X, then the character satisfies $\chi(q, X, L^k) \equiv 0$ (cf. [Ha78] and Appendix III).

Example: For $N = 2$, $c_1(X) \equiv 0$ (2) is precisely the condition $w_2(X) = 0$, which means X has a spin structure. Then $\hat{A}(X) = \chi(X, L)$ where $K = L^2$, so L is a root of the canonical bundle of X. Now the theorem of Atiyah-Hirzebruch (cf. [AtHi70]) states that the character associated to $\hat{A}(X)$ is zero if S^1 acts non-trivially on X.

Now we shall observe that the values at the cusps vanish on the projective spaces of dimension $rN - 1$, $r > 0$:

If $X = P_d(\mathbb{C})$ is a complex projective space and E is the line bundle determined by the divisor $P_{d-1}(\mathbb{C}) \subset P_d(\mathbb{C})$ (so that $E = H^*$), then the canonical bundle is $K = E^{-(d+1)}$. As a special case of a simple residue calculation we obtain (cf. [Hi54]):

$$\chi(P_d(\mathbb{C}), E^i) = \binom{i+d}{d} = \frac{(i+1)\cdots(i+d)}{d!}.$$

This is a polynomial of degree d in i with the zeroes $-1, \ldots, -d$. For $d = rN - 1$, so that $N = (d+1)/r$, we then have

$$\chi(P_d(\mathbb{C}), K^{k/N}) = \chi(P_d(\mathbb{C}), E^{-kr}) = 0 \quad \text{for all } 0 < k < N.$$

In addition, we have (cf. section 5.4):

$$\chi_y(P_{rN-1}(\mathbb{C})) = 1 - y \pm \cdots + (-y)^{rN-1}$$
$$= \frac{1 - (-y)^{rN}}{1 - (-y)},$$

since the $h^{p,q}$ are all zero for $p \neq q$, i.e. $h^{p,p}$ is the $2p$-th Betti number, so is one. From this it follows that

$$\chi_y(P_{rN-1}(\mathbb{C})) = 0 \quad \text{for } -y = e^{2\pi i l/N}, 0 < l < N.$$

Theorem: *The elliptic genus vanishes (identically as a modular form) for complex projective spaces of dimension $rN - 1$; indeed it vanishes for all manifolds which are total space of a bundle with fibre $P_{rN-1}(\mathbb{C})$.*

Proof: The correct proof follows the same lines as for the corresponding theorem about elliptic genera of level two in section 4.6 (use the fact that $f(x - x_1) \cdots f(x - x_{rN})$ is elliptic for $L = 2\pi i(\mathbb{Z}\tau + \mathbb{Z})$).   ☐

The vanishing on $P_{rN-1}(\mathbb{C})$ was to be expected, since as we have seen all values at the cusps vanish for $P_{rN-1}(\mathbb{C})$. We therefore have a cusp form, but one does not come by cusp forms so cheaply.

For a result on multiplicativity in fibre bundles with fibre having first Chern class divisible by N we refer to Appendix III.

7.4 The loop space and the expansion at a cusp

We shall now attempt to define the expression $\chi_y(q, \mathcal{L}X)$ for the loop space $\mathcal{L}X$. We shall proceed by analogy with the case of elliptic genera of level two. On the loop space $\mathcal{L}X$ there is a canonical S^1-action having as fixed point set $(\mathcal{L}X)^{S^1}$ precisely $X \subset \mathcal{L}X$, the space of constant loops. A tangent vector to a constant loop $p \in X$ is a loop in T_pX. This can be regarded as a periodic mapping of $\mathbb{R}$ into a finite dimensional complex vector space, and so has a Fourier expansion with coefficients in T_pX. We obtain the splitting

$$T_p\mathcal{L}X = \sum_{n=-\infty}^{\infty} q^n T_pX,$$

where $q \in S^1$ acts on the n-th summand as multiplication by q^n.

We now apply the equivariant Atiyah-Singer index theorem formally. The χ_y-genus is the index of the Dolbeault complex, computed as the sum of local contributions which are determined by the normal bundle to the fixed components. Hence we should have the formula:

$$\chi_y(q, \mathcal{L}X) = \left(\prod_{i=1}^{d} x_i \left(\prod_{n=-\infty}^{\infty} \frac{1 + yq^n e^{-x_i}}{1 - q^n e^{-x_i}} \right) \right)[X].$$

For $q = 0$ this expression certainly yields $\chi_y(X)$, but since $\sum_{n=-\infty}^{\infty} q^n$ does not converge there is evidently a serious problem with convergence. The product cannot even be multiplied out into a formal power series. We therefore separate the factors for positive and negative n, multiplying the latter by $q^n e^{x_i}$ to obtain:

$$\chi_y(q, \mathcal{L}X) = \left(\prod_{i=1}^{d} x_i \cdot \frac{1 + ye^{-x_i}}{1 - e^{-x_i}} \left(\prod_{n=1}^{\infty} \frac{(1 + yq^n e^{-x_i})(q^n e^{x_i} + y)}{(1 - q^n e^{-x_i})(q^n e^{x_i} - 1)} \right) \right)[X].$$

For $q \in S^1$ this expression still fails to converge; instead we shall consider $q = e^{2\pi i \tau}$, $\tau \in \mathfrak{h}$. Then $\mathrm{Im}\,(\tau) > 0$, hence $|q| < 1$, and the product does indeed converge. Since for the values at the cusps $-y$ is always an N-th root of unity, we shall also consider $-y = \eta$ with $\eta^N = 1$:

$$\chi_{-\eta}(q, \mathcal{L}X) = \left(\prod_{i=1}^{d} x_i \cdot \frac{1 - \eta e^{-x_i}}{1 - e^{-x_i}} \left(\prod_{n=1}^{\infty} \frac{(1 - \eta q^n e^{-x_i})(1 - \eta^{-1} q^n e^{x_i})\eta}{(1 - q^n e^{-x_i})(1 - q^n e^{x_i})} \right) \right)[X].$$

To cancel the η-factor we multiply formally by $\eta^{-d \cdot \sum_{n \geq 1} 1} = \eta^{d/2}$ ($\zeta(0) = -1/2$). If in addition $d \equiv 0 \, (2N)$ so that $\eta^{d/2} = 1$, with $\eta = e^{\alpha}$ we obtain:

$$\chi_{-\eta}(q, \mathcal{L}X) = \left(\prod_{i=1}^{d} x_i e^{\alpha/2} \frac{\sinh\left((x_i - \alpha)/2\right)}{\sinh\left(x_i/2\right)} \right.$$
$$\left. \cdot \left(\prod_{n=1}^{\infty} \frac{\left(1 - q^n e^{-(x_i - \alpha)}\right)\left(1 - q^n e^{(x_i - \alpha)}\right)}{(1 - q^n e^{-x_i})(1 - q^n e^{x_i})} \right) \right)[X].$$

We can write this more simply in terms of the Φ-function. The Φ-function is defined by

$$\Phi(x) = 2 \cdot \sinh(x/2) \cdot \prod_{n=1}^{\infty} \frac{(1 - q^n e^x)(1 - q^n e^{-x})}{(1 - q^n)^2}.$$

For $|q| < 1$, this product yields a holomorphic function in x with zeroes of order 1 at $2\pi i(\mathbb{Z}\tau + \mathbb{Z})$. We have

$$\chi_y(q, \mathcal{L}X) = \left(\prod_{i=1}^{d} x_i e^{\alpha/2} \frac{\Phi(x_i - \alpha)}{\Phi(x_i)} \right)[X]$$

with $-y = e^\alpha$, $(-y)^N = 1$.

For the elliptic genus of level N we have as Fourier expansion (see Appendix I, Theorem 6.4)

$$f(x) = \frac{\Phi(x)\Phi(-\alpha)}{\Phi(x - \alpha)}.$$

The normalization factor $e^{d\alpha/2} \cdot \Phi(-\alpha)^d$ is a modular form of weight d on $\Gamma_1(N)$ when $d \equiv 0\ (N)$ (for N odd) resp. $d \equiv 0\ (2N)$ (for N even). Hence for $(-y)^N = 1$, $\chi_y(q, \mathcal{L}X)$ is a modular function for $\Gamma_1(N)$ under these assumptions.

7.5 The differential equation

The function $h = f^N$ is elliptic with respect to a lattice L. Hence also $f'/f = \frac{1}{N} \cdot h'/h$ is elliptic with respect to L. This function has precisely two poles, namely a pole of order one at the origin as well as at the chosen N-division point P. Hence f'/f defines a double covering of $P_1(\mathbb{C})$ by $\mathbb{C}/L$. On the curve $\mathbb{C}/L$ there is the involution $\tau : x \mapsto -x + P$. The function $f(x)f(-x + P)$ is elliptic with respect to L (under translation by lattice vectors inverse factors from $f(x)$ and $f(-x + P)$ cancel) and has no poles or zeroes, hence it is constant:

$$f(x)f(-x + P) \equiv f(P/2)^2 =: c^{-2} \neq 0.$$

Here $P/2$ is any of the four points $\tilde{P}$ in $\mathbb{C}/L$ with $2\tilde{P} = P$. Logarithmic differentiation then yields:

$$\frac{f'(x)}{f(x)} - \frac{f'(-x + P)}{f(-x + P)} = 0.$$

Hence f'/f is invariant under the involution τ. We thereby obtain an isomorphism $f'/f : (\mathbb{C}/L)/\tau \to P_1(\mathbb{C})$, so that each function on $(\mathbb{C}/L)/\tau$ is a rational function of f'/f. Now the function

$$\frac{1}{f(x)^N} + \frac{1}{f(-x+P)^N} = \frac{1}{f(x)^N} + c^{2N} f(x)^N$$

is certainly invariant under τ and also is elliptic with respect to L, since f^N is; hence it is a rational function of $\xi = f'/f$. Furthermore it has only poles at the origin, and at the point P, both of order N. Since the pole at the origin has leading coefficient 1, while ξ has a simple pole with residue 1 there, we even have an equation of the form

$$\frac{1}{f^N} + a_{2N} \cdot f^N = \xi^N + a_1 \cdot \xi^{N-1} + \cdots + a_{N-1} \cdot \xi + a_N,$$

for we can choose the coefficients $a_1,\dots,a_N$ so that both sides have the same principal part at the origin. Since they are invariant under τ, they also have the same principal parts at P; since they are both elliptic with respect to L, they therefore coincide. Here we have put $a_{2N} = c^{2N}$, so that $a_{2N} \neq 0$. The coefficients $a_1,\dots,a_N$ and a_{2N} can be determined directly from the coefficients of ξ. Hence they are modular forms (on $\Gamma_1(N)$ if $P = 2\pi i/N$) of weight equal to their index (cf. Appendix I, §7).

We shall denote the polynomial of degree N on the right side by P_N. We claim that in this polynomial the coefficient a_{N-1} is equal to zero. Equivalently, we must show that the derivative of P_N has a zero at the origin. We differentiate the above differential equation and obtain

$$P_N'(\xi) \cdot \xi' = \xi \cdot N \cdot \left(a_{2N} \cdot f^N - \frac{1}{f^N}\right).$$

Since the function ξ is elliptic with respect to L but is non-constant, there is a point in $\mathbb{C}/L$ at which it vanishes to an order e. Since the factor in the parentheses on the right only has a pole where $\xi = f'/f$ does not vanish, the order of the zero at this point for the right side of the equation is at least e, since ξ' only vanishes to order $e-1$, we conclude that $P_N'(0) = 0$.

Example: For $N = 2$ the differential equation has the following form:

$$f^{-2} + a_4 \cdot f^2 = (f'/f)^2 + a_2$$
$$\Rightarrow \quad f'^2 = 1 - a_2 \cdot f^2 + a_4 \cdot f^4.$$

With $a_2 = 2\delta$ and $a_4 = \varepsilon$ this is the well-known differential equation for elliptic genera of level two. For arbitrary ε and δ with $\varepsilon \neq 0$ and $\varepsilon - \delta^2 \neq 0$ this determines an elliptic function.

Each differential equation as above yields a unique (up to normalization) power series $f(x)$. For arbitrary coefficients a_r, $f(x)$ will not be an elliptic function. Our a_r

were in fact modular forms of weight r, depending on a lattice with N-division point; evidently they must satisfy several relations, since there are not so many algebraically independent modular forms. What relations must the a_r for an arbitrary N satisfy, in order that one obtain an elliptic function?

The function $\eta = (cf)^{-1} + cf$ is not elliptic for L, but rather for a sub-lattice $\widetilde{L}$ of index N in L. However the involution τ exists on $\mathbb{C}/\widetilde{L}$ as well as on $\mathbb{C}/L$, and η is likewise invariant under τ. If one divides both these curves by the involution, then both become rational curves, i.e. isomorphic to $P_1(\mathbb{C})$. The function

$$\eta : P_1(\mathbb{C}) \cong (\mathbb{C}/\widetilde{L})/\tau \to (\mathbb{C}/L)/\tau \cong P_1(\mathbb{C})$$

is an N-fold covering. It defines the extension of degree N of the function field on $(\mathbb{C}/L)/\tau$ to the function field of $(\mathbb{C}/\widetilde{L})/\tau$. In particular, as can also be established on elementary algebraic grounds, $(cf)^{-N} + (cf)^{N}$ is a polynomial of degree N in $\eta = (cf)^{-1} + cf$; we denote this polynomial by $\widetilde{T}_N$.

Example: The first polynomials $\widetilde{T}_N$ are:

$$z^{-2} + z^2 = \left(z^{-1} + z\right)^2 - 2 \quad \Rightarrow \quad \widetilde{T}_2(x) = x^2 - 2,$$
$$z^{-3} + z^3 = \left(z^{-1} + z\right)^3 - 3\left(z^{-1} + z\right) \quad \Rightarrow \quad \widetilde{T}_3(x) = x^3 - 3x,$$
$$z^{-4} + z^4 = \left(z^{-1} + z\right)^4 - 4\left(z^{-1} + z\right)^2 + 2 \quad \Rightarrow \quad \widetilde{T}_4(x) = x^4 - 4x^2 + 2.$$

We therefore obtain, from the above differential equation, the equation

$$\widetilde{T}_N(\eta) = c^{-N} \cdot P_N(\xi).$$

The curve in the (x,y)-plane defined by the equation $\widetilde{T}_N(x) = c^{-N} \cdot P_N(y)$ is parametrized by ξ and η, hence is isomorphic to $(\mathbb{C}/\widetilde{L})/\tau$ and so is rational. Since it is given by an equation of degree N, it must have a large number of singularities. A point (x,y) on this curve is singular if and only if x and y are critical points of $c^{-N}P_N$ resp. $\widetilde{T}_N$ and the critical values coincide.

The critical points and values of $\widetilde{T}_N$ are easily obtained using the so-called Chebyshev polynomials T_N. These are defined by the equation

$$\cos(N\theta) = T_N(\cos(\theta)).$$

Here is a list of the first few Chebyshev polynomials:

$$T_1(x) = x, \qquad\qquad T_2(x) = 2x^2 - 1,$$
$$T_3(x) = 4x^3 - 3x, \qquad T_4(x) = 8x^4 - 8x^2 + 1.$$

The Chebyshev polynomials always start with $T_N = 2^{N-1}x^N + \cdots$. Since $\cos(\theta) = \left(e^{i\theta} + e^{-i\theta}\right)/2$ and

$$\widetilde{T}_N(2 \cdot \cos(\theta)) = \widetilde{T}_N\left(e^{i\theta} + e^{-i\theta}\right) = e^{iN\theta} + e^{-iN\theta} = 2 \cdot \cos(N\theta),$$

we obtain the relation

$$\widetilde{T}_N(x) = 2 \cdot T_N(x/2).$$

From the defining equation of T_N we see, that the zeroes of T_N' are $x_j = \cos(\pi j/N)$ with $0 < j < N$. At these $N - 1$ points, T_N has local maxima or minima, all with the values ± 1; more precisely:

$$N \text{ even} \quad \Rightarrow \quad (N - 2)/2 \text{ times value } 1,\ N/2 \text{ times value } -1,$$
$$N \text{ odd} \quad \Rightarrow \quad (N - 1)/2 \text{ times value } 1,\ (N - 1)/2 \text{ times value } -1.$$

For $N = 8$ we obtain the following figure:

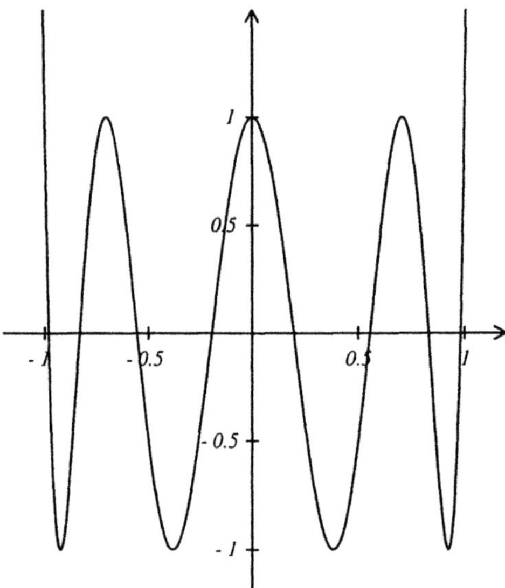

Chebyshev polynomial T_8

For the polynomials $\widetilde{T}_N(x) = 2 \cdot T_N(x/2)$ the $N - 1$ critical points are $\widetilde{x}_j = 2\cos(\pi j/N)$ with $0 < j < N$ and the critical values are ± 2; the distribution of signs is the same as for the Chebyshev polynomials.

So we expect that most of the critical values of the polynomials P_N arising in our differential equation are equal to $\pm 2c^N$. The following discussion will give a more precise result.

The function ξ'^2 is elliptic with respect to L and invariant under the involution τ. It has a pole of order four at the origin and at the point P and its Laurent series at the origin starts with $\frac{1}{x^4} + \cdots$. Therefore there is as above a normalized polynomial $Q(x) = x^4 + q_1 \cdot x^3 + q_2 \cdot x^2 + q_3 \cdot x + q_4$ with

$$Q(\xi) = \xi'^2.$$

The curve given by the equation $y^2 = Q(x)$ is therefore parametrized by $\mathbb{C}/L$, by means of ξ and ξ'. The function ξ assumes each value twice on $\mathbb{C}/L$ (at points which are exchanged with one another by the involution τ); its derivative ξ' has different signs at the corresponding points. Therefore this parametrization is an isomorphism apart from the poles of ξ (at 0 and P).

There is a close relation between the polynomials Q and P_N:

$$P_N(\xi) = f^{-N} + a_{2N} \cdot f^N$$
$$\Rightarrow \quad P_N'(\xi) \cdot \xi' = N \cdot \xi \cdot \left(-f^{-N} + a_{2N} \cdot f^N\right)$$
$$\Rightarrow \quad P_N'(\xi)^2 \cdot \xi'^2 = N^2 \cdot \xi^2 \cdot \left(\left(f^{-N} + a_{2N} \cdot f^N\right)^2 - 4a_{2N}\right)$$
$$\Rightarrow \quad Q(x) \cdot \left(P_N'(x)/x\right)^2 = N^2 \cdot \left(P_N(x)^2 - 4a_{2N}\right).$$

As one easily sees from this equation, away from the origin P_N has only critical points of order precisely two with critical values $\pm\sqrt{4a_{2N}} = \pm 2c^N$. So all the polynomials P_N occurring in the differential equation for f are so-called almost-Chebyshev polynomials.[†]

Definition: *A polynomial $P_N(x) = x^N + \cdots$ is called a **normalized almost-Chebyshev polynomial** if $P_N'(x) = N \cdot x \cdot (x - \xi_1) \cdots (x - \xi_{N-2})$ and the critical values $P_N(\xi_1), \ldots, P_N(\xi_{N-2})$ are equal up to sign.*

7.6 The modular curve

The existence of the equation

$$Q \cdot (P_N'/x)^2 = N^2 \left(P_N^2 - 4a_{2N}\right)$$

relating the polynomials P_N and Q gives us conditions for their coefficients. By differentiating it we obtain:

$$Q' \cdot (P_N'/x)^2 + 2Q \cdot (P_N'/x) \cdot (P_N'' \cdot x - P_N')/x^2 - 2N^2 P_N P_N' = 0$$
$$\Rightarrow \quad Q' \cdot (P_N'/x) + 2Q \cdot (P_N'' \cdot x - P_N')/x^2 - 2N^2 P_N \cdot x = 0$$

(here both P_N'/x and $(P_N'' \cdot x - P_N')/x^2$ are polynomials in x!).

This yields $N + 2$ equations for the coefficients q_1, q_2, q_3, q_4 of Q and $a_1, \ldots, a_N$ of P_N. Since the polynomial Q is monic, the equation of highest degree is always satisfied. The next equations successively determine the a_r as homogeneous polynomials of weight

---

[†]   F.H.: This terminology was used during my course. I did not know that these polynomials exist in the literature under the name Zolotarev polynomials. I wrote to Prof. Rivlin about the almost-Chebyshev polynomials and got from him the information on Zolotarev polynomials.

r in the $q_1,\ldots,q_4$ (of weights 1 to 4). Since for our construction of the elliptic genus $a_{N-1} = 0$, one obtains a homogeneous equation of weight $N-1$ in $q_1,\cdots,q_4$. There still remains the equation for the constant term, which yields a second condition of weight $N+1$ on q_1 to q_4. These two equations determine those polynomials Q for which a polynomial P_N with $Q \cdot (P_N'/x)^2 = N^2 \cdot (P_N^2 - 4a_{2N})$ exists (with $a_{2N} = (a_N/2)^2 - q_4(a_{N-2}/N)^2$).

Example: For $N = 2$ we have

$$\left(4x^3 + 3q_1 x^2 + 2q_2 x + q_3\right) \cdot 2 + 2Q \cdot 0 - 8x\left(x^2 + a_2\right) = 0 \quad \text{and}$$
$$a_4 = (a_2/2)^2 - q_4(1/2)^4.$$

From these equations we get

$$q_1 = 0, \qquad q_3 = 0,$$
$$a_2 = q_2/2, \qquad a_4 = q_2^2/16 - q_4/4.$$

With $a_2 = 2\delta$ and $a_4 = \varepsilon$ we have

$$P_2(x) = x^2 + 2\delta \quad \text{and}$$
$$Q(x) = x^4 + 4\delta \cdot x^2 + 4\left(\delta^2 - \varepsilon\right).$$

At the signature cusp we have $\delta^2 = \varepsilon$. Hence Q has a double zero at the origin in this case. At the $\hat{A}$ cusp we have $\varepsilon = 0$, hence Q has two double zeroes.

For $N = 3$ we have:

$$\left(4x^3 + 3q_1 x^2 + 2q_2 x + q_3\right)(3x + 2a_1)$$
$$+6\left(x^4 + q_1 x^3 + q_2 x^2 + q_3 x + q_4\right) - 18\left(x^4 + a_1 x^3 + a_3 x\right) = 0$$

In this case we get the equations

$$a_1 = 3/2 \cdot q_1, \qquad\qquad q_2 = -3/4 \cdot q_1^2,$$
$$a_2 = \left(2 \cdot q_3 - q_1^3\right)/4, \qquad q_4 = -q_1 q_3/2,$$
$$a_6 = \left(2 \cdot q_3 - q_1^3\right)^2/64.$$

The two homogeneous equations satisfied by the coefficients q_1 through q_4 determine a curve C_N in the weighted projective space with the weighted homogeneous coordinates q_1 through q_4, q_i being of weight i. Each elliptic curve with distinguished primitive N-division point gives rise to a point on this modular curve C_N. If one parametrizes such elliptic curves by $\tau \in \Gamma_1(N)\backslash\mathfrak{h}$ (cf. Appendix I, §7), then this mapping is simply given by $(q_1(\tau) : \ldots : q_4(\tau))$; this is well-defined since q_i is a modular form of weight i and the coordinates are weighted homogeneous. The modular curve is in general not irreducible, but the extension of the mapping to $\overline{\Gamma_1(N)\backslash\mathfrak{h}}$ is surjective on individual components for dimensional reasons, which then parametrize our lattices.

Now each elliptic curve with an N-division point is also a curve with an $N \cdot d$-division point. What happens to the polynomials P_N, and how does the embedding ι of the corresponding modular curves look?

We consider the differential equation

$$c^{-N} \cdot P_N(\xi) = \tilde{T}_N(\eta).$$

In order to bring it to the form

$$c^{-dN} \cdot P_{dN}(\xi) = \tilde{T}_{dN}(\eta),$$

we must apply the polynomial $\tilde{T}_d$ (it follows immediately from the definition of the Chebyshev polynomials that $\tilde{T}_d(\tilde{T}_N) = \tilde{T}_{dN}$). We therefore obtain

$$P_{dN} = c^{dN} \cdot \tilde{T}_d(c^{-N} \cdot P_N).$$

This polynomial P_{dN} is again almost-Chebyshev, and satisfies

$$Q \cdot (P'_{dN}/x)^2 = N^2 \cdot (P^2_{dN} - 4c^{2dN})$$

with the same polynomial Q as in the corresponding equation for P_N. For the polynomial Q it is immediately clear that it does not change when an N-division point is re-interpreted as a dN-division point, i.e. on the modular curve C_N the embedding into C_{dN} is simply the inclusion of some components.

Example: For $N = 4$ the equations for Q are given by

$$q_3 = -\frac{3}{8} \cdot q_1 (q_1^2 + 4q_2),$$

$$q_1 q_4 = \frac{1}{64} \cdot q_1 (4q_2 + 5q_1^2)(4q_2 + q_1^2).$$

These equations encompass the case $N = 2$ ($q_1, q_3 = 0$). For $q_1 \neq 0$, q_1 and q_2 are arbitrary while q_3, q_4 depend on q_1, q_2.

The modular curve C_N has in general many irreducible components; those which belong to an elliptic curve with a primitive N-division point are not in the image of an embedding as above. Since for a prime $N = p$ all N-division points are primitive, this should be the entire modular curve. Hence for a prime there should be only one component.

The map from $\Gamma_1(N)\backslash \mathfrak{h}$ to C_N can be extended to the cusps of $\Gamma_1(N)$. What does the differential equation at the cusps look like?

One of the values of the genus at the cusps is the normalized χ_y-genus $\chi_y(X)/(1+y)^{\dim X}$, where $-y$ must be an N-th root of unity different from 1.

Hence the Todd genus $\chi_0(X)$ never arises. However, this is the only genus which is invariant under all birational transformations. For we have

$$\chi_0(X) = \sum_{q=0}^{\dim X} (-1)^q h^{0,q} = \sum_{q=0}^{\dim X} (-1)^q h^{q,0}$$

when X is Kähler, and the numbers $h^{q,0}$ are birational invariants (including the geometric genus $h^{\dim X,0}$). Hence χ_0 is invariant under birational transformations. Since in the cobordism ring $\Omega \otimes \mathbb{Q}$ each manifold X is equal to a sum of products of projective spaces, which in turn is birationally equivalent to a multiple of $P_1(\mathbb{C})^{\dim X}$, each genus with this property is determined by its value on $P_1(\mathbb{C})$, hence is a multiple of χ_0.

The power series

$$f(x) = \frac{1+y}{1+ye^{-x}} \cdot \left(1 - e^{-x}\right)$$

belonging to the normalized χ_y-genus satisfies the differential equation

$$f'(x) = \left(1 - \frac{1}{1+y}f(x)\right) \cdot \left(1 + \frac{y}{1+y}f(x)\right).$$

If we bring the differential equation to our form, then with $c = \frac{\sqrt{-y}}{1+y}$ we have:

$$c^{-1} \cdot \left(\frac{f'}{f} + \frac{1-y}{1+y}\right) = (cf)^{-1} + cf$$

$$\Rightarrow \quad P_1(x) = x + \frac{1-y}{1+y}.$$

But in this differential equation the coefficient of f'/f is non-zero.

We return to the embedding ι of modular curves. What follows for it from our condition $a_{N-1} = 0$? The formula for P_{dN} yields

$$P'_{dN}(0) = c^{dN} \cdot \tilde{T}'_d\left(c^{-N} \cdot P_N(0)\right) \cdot c^{-N} \cdot P'_N(0).$$

So if $P'_N(0)$ is zero, this remains true for $P'_{dN}(0)$; if $P'_N(0)$ is not zero, then $c^{-N} \cdot P_N(0)$ must be a zero of $\tilde{T}'_d$. In the differential equation of the normalized χ_y-genus we have $N = 1$ and $c = \frac{\sqrt{-y}}{1+y}$. This forces $-y$ to be a d-th root of unity (check this!). Then we have

$$P_d = c^d \tilde{T}_d\left(c^{-1}\left(x + \frac{1-y}{1+y}\right)\right),$$

i.e. P_d is essentially a Chebyshev polynomial. Hence the value at the critical point 0 is the same as at the other critical points, $a_d^2 = 4a_{2d}$. In addition, we find the polynomial Q to be

$$Q(x) = x^2\left(\left(x + \frac{1-y}{1+y}\right)^2 - 4c^2\right) = x^2\left(x^2 + 2\frac{1-y}{1+y}x + 1\right);$$

in particular, we have q_3, $q_4 = 0$, i.e. Q has a double zero at $x = 0$. We have therefore handled one kind of cusp.

For the derivation of the differential equation we have always assumed that $c \neq 0$, hence also $a_{2N} \neq 0$. If now $a_{2N} = 0$, i.e. apart from the origin only the critical value zero appears, then it follows by taking into account the other conditions on the coefficients a_r that the N-fold covering has only two branch points and the roots are rational. The genera which now arise are $\chi(X, K^{k/N})$ for $0 < k < N$. The corresponding power series is

$$Q(x) = \frac{x}{1 - e^{-x}} \cdot e^{-\frac{k}{N}x}$$
$$\Rightarrow \quad f(x) = \left(1 - e^{-x}\right) \cdot e^{\frac{k}{N}x}.$$

What differential equation does $f(x) = \left(1 - e^{-x}\right) \cdot e^{\frac{k}{N}x}$ satisfy?

$$e^{\frac{k}{N}x} = \frac{(N-k) \cdot f + Nf'}{N}, \quad e^{-\frac{N-k}{N}x} = \frac{-kf + Nf'}{N},$$
$$\left(e^{\frac{k}{N}x}\right)^{N-k} \cdot \left(e^{-\frac{N-k}{N}x}\right)^k = 1$$
$$\Rightarrow \quad \left(\frac{f'}{f} + \frac{N-k}{N}\right)^{N-k} \cdot \left(\frac{f'}{f} - \frac{k}{N}\right)^k = \frac{1}{f^N}.$$

One easily concludes that in fact we have $a_{N-1} = 0$. Apart from the origin there are now two critical points, an $(N - k - 1)$-fold one at k/N and a $(k - 1)$-fold one at $k/N - 1$, both with the critical value 0. The polynomial Q is then

$$Q(x) = (x - k/N)^2 \cdot (x - (k/N - 1))^2,$$

so it has two double zeroes. In this case, the functions f'/f and f parametrize an N-sheeted covering of the Riemann sphere having these two branch points.

We have therefore found an interpretation of the modular curve which parametrizes elliptic curves with a non-zero N-division point by almost-Chebyshev polynomials (for a more detailed discussion of this relation between elliptic curves and almost-Chebyshev polynomials we refer to Appendix IV). If $N = p$ is a prime, then this curve has only one component. The entire family of elliptic curves then forms an elliptic surface, which is known as a Shioda surface. Over each point of the modular curve lies an elliptic curve in the surface; over the points of the modular curve corresponding to the cusps of $\Gamma_1(N)$

the elliptic curves degenerate. For $p > 3$, the $\chi(X, K^{k/p})$-genus yields a rational curve with a double point, the double point yielding precisely the two branch points. For the χ_y-genus the elliptic curve degenerates to a cycle of p rational curves. These possess a natural involution. Dividing by the involution, we then have $(p+1)/2$ components. These are precisely represented by the genera. For $-y$ is permitted to run through all primitive p-th roots of unity, but y and $1/y$ yield the same genus.

Exercise: Calculate the Euler number of this elliptic surface.

8 A divisibility theorem for elliptic genera

8.1 The theorem of Ochanine

In this chapter we will use the (cf. [AtHi59b])

Theorem: *Let W be the complex extension of a real vector bundle (hence $W \cong W^*$) over a compact, oriented, differentiable spin manifold M with $\dim M \equiv 4$ (8). Then $\hat{A}(M, W) \in 2\mathbb{Z}$.*

Remark: For a complex manifold X of dimension d, we have

$$\hat{A}(X, W) = \chi(X, K^{1/2} \otimes W) = \sum_{i=0}^{d} (-1)^i \dim H^i(X; K^{1/2} \otimes W).$$

We may transform these summands by Serre duality ($\dim X = d$):

$$\begin{aligned}
H^i(X; K^{1/2} \otimes W) &\cong H^{d-i}(X; K \otimes K^{-1/2} \otimes W^*) \\
&\cong H^{d-i}(X; K^{1/2} \otimes W^*) \\
&\cong H^{d-i}(X; K^{1/2} \otimes W).
\end{aligned}$$

We therefore see that modulo 2 all summands apart from the middle one ($i = d/2$) cancel. Now Serre duality is given by a bilinear form, which is skew-symmetric for odd $i = d/2$. Therefore $H^{d/2}(X; K^{1/2} \otimes W)$ must have an even dimension.

For the proof of the theorem in the oriented case, one uses the Dirac operator and the Atiyah-Singer index theorem.

Now we want to discuss the following (cf. [Oc81])

Theorem (Ochanine): *Let M be a compact, oriented, differentiable spin manifold with $\dim M \equiv 4$ (8). Then we have:*

$$\operatorname{sign}(M) \equiv 0 \ (16).$$

We shall give the proof of the theorem in the next section.

Remark: Since in dimension four only the Pontrjagin number p_1 exists, in this case all genera differ from one another by a constant factor; in particular (cf. section 6.4)

$$\operatorname{sign}(M) = -8 \cdot \hat{A}(M).$$

Using the above theorem on twisted $\hat{A}$-genera (with W a trivial line bundle), we obtain the theorem of Ochanine in dimension four. Rohlin proved this by different methods in 1950. The result is sharp in dimension four, since the $K3$-surface has signature 16.

For $k \geq 2$ we obtain by the plumbing construction a topological manifold M^{4k} with intersection form E_8. The manifold therefore has signature 8. In addition it is $(2k-1)$-connected, hence all its cohomology groups vanish in dimensions below $2k$. Following Thom and Wu, the Stiefel-Whitney classes can also be defined for topological manifolds (cf. [AtHi61a]). In our case, we have $w_1(M) = w_2(M) = 0$, since there is no cohomology in dimensions 1 and 2. If this manifold had a differentiable structure, all conditions of the above theorem would be satisfied, hence the signature would be at least 16. We therefore conclude that M has no differentiable structure, and on the other hand that the theorem of Ochanine actually makes use of the differentiable structure. As we have seen, the theorem cannot be rescued for topological manifolds. We shall nevertheless investigate this case further. On the cohomology ring of a topological manifold M of dimension n there exist natural cohomology operations (cf. [StEp62]), the so-called Steenrod squares, with:

1)	$Sq^i : H^j(M; \mathbb{Z}/2\mathbb{Z}) \to H^{j+i}(M; \mathbb{Z}/2\mathbb{Z})$,
2)	$Sq^0 = \mathrm{id}$,
3)	$Sq^i(x) = x^2$ for $x \in H^i(M; \mathbb{Z}/2\mathbb{Z})$,
4)	$Sq^i(x) = 0$ if $x \in H^j(M; \mathbb{Z}/2\mathbb{Z})$ with $j < i$,
5)	$Sq^i(xy) = \sum_{j=0}^{i} Sq^j(x) \cdot Sq^{i-j}(y)$.

In addition there is a cohomology class $U_i \in H^i(M; \mathbb{Z}/2\mathbb{Z})$ with

$$Sq^i(x) = U_i \cdot x \quad \text{for } x \in H^{n-i}(M; \mathbb{Z}/2\mathbb{Z}).$$

The Stiefel-Whitney classes can also be defined for topological manifolds, and we then have:
$$w = Sq(U), \text{ i.e.}$$
$$1 + w_1 + \cdots = Sq^0(1 + U_1 + \cdots) + Sq^1(1 + U_1 + \cdots) + \cdots.$$

For the first three classes this means:
$$w_1 = Sq^1(1) + Sq^0(U_1) = U_1,$$
$$w_2 = Sq^2(1) + Sq^1(U_1) + Sq^0(U_2) = U_1^2 + U_2,$$
$$w_3 = Sq^3(1) + Sq^2(U_1) + Sq^1(U_2) + Sq^0(U_3)$$
$$= Sq^1(U_2) + U_3.$$

More is implied by our assumptions $w_1, w_2 = 0$. For one obtains the odd Stiefel-Whitney classes w_{2i+1} as the modulo 2 reduction of the image of w_{2i} under the Bockstein operator for the coefficient sequence

$$0 \to \mathbb{Z} \xrightarrow{\cdot 2} \mathbb{Z} \to \mathbb{Z}/2\mathbb{Z} \to 0.$$

Since we have $w_2 = 0$, it follows that also $w_3 = 0$. As we have already seen, these three Stiefel-Whitney classes are zero if and only if U_1, U_2 and U_3 vanish, that is the Steenrod squares Sq^1, Sq^2 and Sq^3 are identically zero in codimensions 1, 2 and 3, resp. We shall now exploit this in order to learn something about the intersection form of the manifold, which is defined on the integral lattice of the cohomology in the middle dimension. For this we make use of the so-called Adem relations which are general relations on the Steenrod squares. Two special cases read (with $i \geq 1$ and coefficients modulo 2):

$$
\begin{aligned}
Sq^2 Sq^{4i} &= \binom{4i-1}{2} \cdot Sq^{4i+2} + \binom{4i-2}{0} \cdot Sq^{4i+2} Sq^1 \\
&= Sq^{4i+2} + Sq^{4i+1} Sq^1, \\
Sq^3 Sq^{4i-1} &= \binom{4i-2}{3} \cdot Sq^{4i+2} + \binom{4i-3}{1} \cdot Sq^{4i+1} Sq^1 \\
&= Sq^{4i+1} Sq^1, \\
\Rightarrow \quad Sq^{4i+2} &= Sq^2 Sq^{4i} + Sq^3 Sq^{4i-1}.
\end{aligned}
$$

The right side is zero when applied to cohomology classes of codimension $4i + 2$, since Sq^2 and Sq^3 vanish in codimension 2 resp. 3. Therefore Sq^{4i+2} is zero in codimension $4i + 2$.

Definition: *The **intersection form** of M^{4n} is called **even** if*

$$
x^2[M] \equiv 0 \quad (2) \quad \text{for all } x \in H^{2n}(M; \mathbb{Z}).
$$

As we have seen above, we obtain for any $8k + 4$-dimensional topological manifold with vanishing w_1 and w_2 that $x \cdot x = Sq^{4k+2}(x) = 0$ for $x \in H^{4k+2}(M; \mathbb{Z}/2\mathbb{Z})$. So our intersection form is even. By Poincaré duality it is always unimodular, i.e. nonsingular. Now the signature of an even unimodular quadratic form is always divisible by 8. For topological manifolds with $w_1 = w_2 = 0$, Ochanine's theorem therefore still holds in the weaker form

$$
\text{sign}\left(M^{8k+4}\right) \equiv 0 \quad (8).
$$

Just as the Stiefel-Whitney classes w_j can be expressed in terms of the U_i by means of the Steenrod squares, one can express the U_i in terms of the w_j, indeed as polynomials in the w_j. We shall consider these polynomials in greater detail (cf. [Wu50]):

Theorem (Wu): *There exists a weighted homogeneous polynomial $U_n(w_1, \ldots, w_n)$ of degree n where w_i has degree i, such that for each compact topological manifold M^{2n} with Stiefel-Whitney classes w_i there holds*

$$
x^2 = (x \cdot U_n(w_1, \ldots, w_n)) \quad \text{for all } x \in H^n(M; \mathbb{Z}/2\mathbb{Z}).
$$

The polynomials U_i form a multiplicative sequence in the Stiefel-Whitney classes, having the characteristic power series

$$1 + \sum_{i=0}^{\infty} x^{2^i} \equiv \frac{2x}{1 - e^{-2x}} \quad (2).$$

The power series on the right side yields precisely the Todd polynomials, up to a factor 2^i in front of x^i. We therefore have

$$U_n \equiv 2^n \cdot T_n(w_1, \ldots, w_n) \quad (2).$$

The factor 2^n is needed to cancel the powers of 2 in the denominator of the coefficients of the power series $x/(1 - e^{-x})$.

Example:

$$T_2 = \frac{c_2 + c_1^2}{12} \quad \Rightarrow \quad U_2 = w_2 + w_1^2,$$

$$T_3 = \frac{c_1 c_2}{24} \quad \Rightarrow \quad U_3 = w_1 w_2,$$

$$T_4 = \frac{-c_4 + c_3 c_1 + 3c_2^2 + 4c_2 c_1^2 - c_1^4}{720} \quad \Rightarrow \quad U_4 = w_4 + w_3 w_1 + w_2^2 + w_1^4.$$

Hence we see that, when $w_1, w_2 = 0$, the polynomial U_2 vanishes. Since in addition $w_3 = 0$, we also have

$$U_6 = w_4 w_2 + w_4 w_1^2 + w_3^2 + w_3 w_2 w_1 + w_3 w_1^3 + w_2^2 w_1^2 = 0.$$

Here we are fortunate, since theoretically w_6 could also appear as a term in U_6. However, the coefficient of c_6 in T_6 is even. As we have seen, multiplication by the polynomial U_{4i+2} on the cohomology in codimension $4i + 2$ of any manifold with $w_1, w_2, w_3 = 0$ is the zero mapping. Since this multiplication originates from a non-degenerate pairing of the corresponding cohomology groups, U_{4i+2} is zero in the cohomology ring of every such manifold. Nevertheless, as a polynomial U_{4i+2} need not vanish modulo the indeterminates w_1, w_2, w_3.

Example:

$$U_{10} \equiv w_5^2 \quad (w_1, w_2, w_3), \qquad U_{14} \equiv w_4 w_5^2 \quad (w_1, w_2, w_3).$$

8.2 Proof of Ochanine's theorem

Having already formulated the theorem in the last section, we shall now prove it. For this we shall recall once more some of the general theory of elliptic genera of level N.

Let $L = 2\pi i(\mathbb{Z}\tau + \mathbb{Z}) \subset \mathbb{C}$ be a lattice, and let h be the elliptic function for this lattice with divisor $(h) = N \cdot (0) - N \cdot (\alpha)$ and the normalization $h(x) = x^N + \cdots$. Here α is an N-division point of the lattice L, which we take as $\alpha = 2\pi i/N$. In addition, we have denoted by f the N-th root of h, normalized by $f(x) = x + \cdots$. For a complex manifold X of complex dimension d, the elliptic genus $\varphi_{N,\alpha}$ of level N is then given by

$$\varphi_{N,\alpha}(X) = \left(\prod_{i=1}^{d} \frac{x_i}{f(x_i)}\right)[X].$$

The power series f can be explicitly determined. With

$$\Phi(\tau, x) = 2 \cdot \sinh\left(x/2\right) \cdot \prod_{n=1}^{\infty} \frac{(1 - q^n e^x)(1 - q^n e^{-x})}{(1 - q^n)^2}$$

we have

$$f(x) = \frac{\Phi(\tau, x)\Phi(\tau, -\alpha)}{\Phi(\tau, x - \alpha)}.$$

The elliptic genus $\varphi_{N,\alpha}(X)$ of X is a modular form of weight d on $\Gamma_1(N)$ ($\Gamma_1(N)$, in view of our choice of α). As the expansion of the elliptic genus at a χ_y-cusp we obtained

$$\chi_y(q, \mathcal{L}X) = \varphi_{N,\alpha}(X) \cdot \Phi(\tau, -\alpha)^d \cdot (-y)^{d/2}$$

with $-y = e^\alpha$. The normalization factor $\left(e^{\alpha/2} \cdot \Phi(\tau, -\alpha)\right)^{-1}$ is a modular form of weight 1 on $\Gamma_1(N)$, up to $2N$-th roots of unity. If $2N$ divides the dimension d, then $\left(e^{\alpha/2} \cdot \Phi(\tau, -\alpha)\right)^{-d}$ is likewise a modular form of weight d (cf. section 7.4). Hence $\chi_y(q, \mathcal{L}X)$ is a modular function on $\Gamma_1(N)$ if $2N \mid d$.

We also have derived a product representation:

$$\chi_y(q, \mathcal{L}X) = \left(\prod_{i=1}^{d} x_i \cdot \frac{1 + ye^{-x_i}}{1 - e^{-x_i}} \left(\prod_{n=1}^{\infty} \frac{1 + yq^n e^{-x_i}}{1 - q^n e^{-x_i}} \cdot \frac{1 + y^{-1}q^n e^{x_i}}{1 - q^n e^{x_i}}\right)\right)[X]$$

$$= \chi_y\left(X, \bigotimes_{n=1}^{\infty} \Lambda_{yq^n} T^* \otimes \bigotimes_{n=1}^{\infty} S_{q^n} T^* \otimes \bigotimes_{n=1}^{\infty} \Lambda_{y^{-1}q^n} T \otimes \bigotimes_{n=1}^{\infty} S_{q^n} T\right)$$

$$= \chi_y(X) + \cdots.$$

The representation in terms of the twisted χ_y-genera follows once again from the formulas in section 1.5. The coefficients of q^i are therefore algebraic integers in the cyclotomic field $\mathbb{Q}[y]$. For $y = 1$ one can combine the terms for T and for T^* and express them using $T_{\mathbb{C}} = T \oplus T^*$. In order to arrive at our genus from this expansion, we must still divide by $\Phi(\tau, -\alpha)^d(-y)^{d/2}$. Now we have

$$
\begin{aligned}
\Phi(\tau, -\alpha)^d &= (2 \cdot \sinh(-\alpha/2) \cdot (\text{integral } q\text{-expansion}))^d \\
&= \left(e^{-\alpha/2} - e^{\alpha/2}\right)^d \cdot (\text{integral } q\text{-expansion}) \\
&= e^{-d\alpha/2}(1 - e^{\alpha})^d \cdot (\text{integral } q\text{-expansion}) \\
&= (-y)^{-d/2}(1 + y)^d \cdot (\text{integral } q\text{-expansion}),
\end{aligned}
$$

where "integral q-expansion" always starts with $1 + O(q)$. Hence it is invertible in $\mathbb{Z}[[q]]$ and we have

$$
\varphi_{N,\alpha}(X) = \frac{(\text{integral } q\text{-expansion})}{(1 + y)^d}.
$$

How does the expansion look at one of the other cusps?

Instead of expanding the genus at another cusp, we can also change the N-division point (cf. Appendix I, §6). We now choose $\alpha = 2\pi i k\tau/N$ with $0 < k < N$. The power series f then takes the form

$$
f(x) = \frac{\Phi(\tau, x)\Phi(\tau, -\alpha)}{\Phi(\tau, x - \alpha)} \cdot e^{(k/N)x}
$$

and for the genus we obtain:

$$
\varphi_{N,\alpha}(X) = \left(\prod_{i=1}^{d} \left(\frac{x_i}{1 - e^{-x_i}} \cdot \frac{1 - q^{k/N}e^{-x_i}}{1 - q^{k/N}} \cdot e^{-(k/N)x_i} \right. \right.
$$
$$
\left. \left. \prod_{n=1}^{\infty} \frac{\left(1 - q^{(n+k/N)}e^{-x_i}\right)\left(1 - q^{(n-k/N)}e^{-x_i}\right)(1 - q^n)^2}{(1 - q^n e^{-x_i})(1 - q^n e^{x_i})\left(1 - q^{(n+k/N)}\right)\left(1 - q^{(n-k/N)}\right)} \right) \right)[X].
$$

In this expression $q^{1/N}$ appears; $\varphi_{N,\alpha}(X)$ is a modular form on $\Gamma^1(N)$. In order to obtain a modular form on $\Gamma_1(N)$ we must replace q by q^N; the q-expansion of this modular form $\widetilde{\varphi}_{N,\alpha}(X)$ has integral coefficients:

$$
\widetilde{\varphi}_{N,\alpha}(X) = (\text{Norm.})
$$
$$
\cdot \chi\left(X, K^{k/N} \otimes \bigotimes_{n \equiv k \, (N)} \Lambda_{q^n}T^* \otimes \bigotimes_{n=1}^{\infty} S_{q^{nN}}T^* \otimes \bigotimes_{n \equiv -k \, (N)} \Lambda_{q^n}T \otimes \bigotimes_{n=1}^{\infty} S_{q^{nN}}T \right).
$$

Here the normalization factor "Norm." has integral coefficients, and the integrality of the coefficients of the Riemann-Roch expression follows from that of $\chi(X, K^{k/N} \otimes W)$ for

$c_1 \equiv 0\ (N)$. Therefore our genus $\varphi_{N,\alpha}$ has an integral q-expansion for $\alpha = 2\pi i k\tau/N$, provided that c_1 is divisible by N.

We now specialize these considerations to the elliptic genera of level two. Let X be an oriented differentiable manifold of dimension $4k$. The elliptic genus of level two then yields a modular form of weight $2k$ on $\Gamma_1(2) = \Gamma_0(2)$. There are two cusps. At one cusp (the signature) the expansion is integral up to the factor 2^{-2k} (for $d = 2k$ and $y = 1$), at the other cusp (the $\hat{A}$-genus) the q-expansion is integral if $c_1 \equiv 0\ (2)$, i.e. $w_2 = 0$. The coefficients of this q-expansion are twisted $\hat{A}$-genera $\hat{A}(X, W)$. The bundles W which occur are complex bundles associated to $T_{\mathbb{C}}$, hence satisfy $W^* \cong W$.

For this genus we began with the differential equation

$$f'^{2} = 1 - 2\delta \cdot f^2 + \varepsilon \cdot f^4.$$

From the product representation of f at the signature cusp one can easily determine the coefficients of $f'(x)/f(x)$, and then use the differential equation to obtain the following formulas for δ and ε (cf. Appendix I, §§3,4):

$$\delta = 1/4 + 6 \sum_{n=1}^{\infty} \sum_{\substack{d|n \\ d \text{ odd}}} d \cdot q^n,$$

$$\varepsilon = 1/16 + \sum_{n=1}^{\infty} \sum_{d|n} (-1)^d d^3 \cdot q^n.$$

If one expands f at the other cusp, one must obviously also do this for δ and ε in the differential equation. We shall denote by a tilde the conjugate expansion for $\Gamma_0(2)$ at the other cusp,

$$\widetilde{f}'^{2} = 1 - 2\widetilde{\delta} \cdot \widetilde{f}^2 + \widetilde{\varepsilon} \cdot \widetilde{f}^4.$$

Under this passage, we must replace τ by $-1/\tau$ and then τ by 2τ, in order to arrive at the modular group $\Gamma_0(2)$. We then have (again, cf. Appendix I, §§3,4):

$$\widetilde{\delta} = -\frac{1}{2} \cdot \delta = -\frac{1}{8} - 3 \cdot \sum_{n=1}^{\infty} \sum_{\substack{d|n \\ d \text{ odd}}} d \cdot q^n,$$

$$\widetilde{\varepsilon} = \frac{\delta^2 - \varepsilon}{4} = \sum_{n=1}^{\infty} \sum_{\substack{d|n \\ n/d \text{ odd}}} d^3 \cdot q^n.$$

Now each modular form on $\Gamma_0(2)$ is a homogeneous polynomial in δ and ε, hence also a polynomial in $8\widetilde{\delta}$ and $\widetilde{\varepsilon}$:

$$\widetilde{\varphi}\left(X^{4k}\right) = \sum_{\substack{a,b \\ 2a+4b=2k}} c_{a,b} \cdot \left(8\widetilde{\delta}\right)^a \cdot \widetilde{\varepsilon}^b = P\left(8\widetilde{\delta}, \widetilde{\varepsilon}\right).$$

If $w_2(X) = 0$, we know that $\widetilde{\varphi}(X)$ has an integral q-expansion, and then the coefficients $c_{a,b}$ must likewise be integral (proceed by induction, noting that $8\widetilde{\delta} = -1 + \cdots$ and $\widetilde{\varepsilon} = q + \cdots$), hence

$$\widetilde{\varphi}(X^{4k}) \in \mathbb{Z}[8\widetilde{\delta}, \widetilde{\varepsilon}]$$
$$\Rightarrow \quad \varphi(X^{4k}) \in \mathbb{Z}[8\delta, \varepsilon].$$

Now at the signature cusp $2^{2k} \cdot \varphi(X)$ is integral,

$$
\begin{aligned}
\operatorname{sign}(q, \mathcal{L}X) &= \varphi(X) \cdot 2^{2k} \cdot (\text{integral } q\text{-expansion}) \\
&= 2^{2k} \cdot P(8\delta, \varepsilon) \cdot (\text{integral } q\text{-expansion}) \\
&= P(32\delta, 16\varepsilon) \cdot (\text{integral } q\text{-expansion}), \\
32\delta &= 8 \cdot (1 + 24q \cdot (\cdots)), \\
16\varepsilon &= 1 + 16q \cdot (\cdots).
\end{aligned}
$$

If our manifold has dimension $4k$ with k odd, then the polynomial P is divisible by 32δ, since otherwise the weight could not be $2k$. Hence all the coefficients of the modular function on the left side are divisible by 8:

$$\operatorname{sign}(q, \mathcal{L}X) \equiv 0 \ (8).$$

We are now just a factor of 2 away from Ochanine's theorem. By the theorem at the start of section 8.1, the expansion at the $\hat{A}$-cusp is not only integral, indeed its coefficients are all even. It follows at once that all the coefficients $c_{a,b}$ of the polynomial P must likewise be even. We conclude that

$$\operatorname{sign}(q, \mathcal{L}X) \equiv 0 \ (16).$$

In particular this holds for the constant term, the signature of X.

Appendix I: Modular forms

by Nils-Peter Skoruppa, Max-Planck-Institut für Mathematik, Bonn

1   Fundamental concepts

Let $\mathfrak{h} = \{\tau \in \mathbb{C} \mid \operatorname{Im}(\tau) > 0\}$ be the upper half-plane of the complex numbers. The group $\mathrm{SL}_2(\mathbb{R})$ acts on $\mathfrak{h}$ by

$$(A, \tau) \mapsto A\tau := \frac{a\tau + b}{c\tau + d} \quad \text{for } A = \begin{pmatrix} a & b \\ c & d \end{pmatrix} \in \mathrm{SL}_2(\mathbb{R}) \text{ and } \tau \in \mathfrak{h}.$$

This action is the restriction to $\mathfrak{h}$ of the natural action of $\mathrm{SL}_2(\mathbb{R})$ on $P_1(\mathbb{C})$, if one regards $\mathfrak{h}$ as a subset of $P_1(\mathbb{C})$ via $\tau \mapsto \left[\begin{smallmatrix} \tau \\ 1 \end{smallmatrix}\right]$. Because $\operatorname{Im}(A\tau) = \operatorname{Im}(\tau)/|c\tau + d|^2$, $\mathfrak{h}$ is in fact invariant under $\mathrm{SL}_2(\mathbb{R})$.

Remark: This action factorizes to an action of the group $\mathrm{PSL}_2(\mathbb{R}) = \mathrm{SL}_2(\mathbb{R})/\{\pm 1\}$; $\mathrm{PSL}_2(\mathbb{R})$ is isomorphic to the group of biholomorphic transformations of $\mathfrak{h}$.

For a subgroup $\Gamma \subset \mathrm{SL}_2(\mathbb{R})$, $\Gamma\backslash\mathfrak{h}$ becomes a topological space by means of the quotient topology. If Γ is a discrete subgroup, then a unique complex structure exists on $\Gamma\backslash\mathfrak{h}$ so that the natural projection $\pi : \mathfrak{h} \to \Gamma\backslash\mathfrak{h}$ is holomorphic.

Each compact Riemann surface of genus greater or equal to 2 can be obtained in that way. For if X is a Riemann surface and $\widetilde{X}$ its universal covering, then by the Riemann mapping theorem $\widetilde{X}$ is isomorphic to either $P_1(\mathbb{C})$, $\mathbb{C}$ or $\mathfrak{h}$; Therefore we have $X \cong P_1(\mathbb{C})$, or $X \cong \mathbb{C}/L$ for a lattice L in $\mathbb{C}$ (and X has genus 1), or $X \cong \Gamma\backslash\mathfrak{h}$ for suitable $\Gamma \subset \mathrm{SL}_2(\mathbb{R})$ (and X has genus greater or equal to 2). In the last two cases L, resp. $P\Gamma := \Gamma/(\{\pm 1\} \cap \Gamma)$ is isomorphic to the group of covering transformations of the covering $\widetilde{X} \to X$, i.e. is isomorphic to $\pi_1(X)$.

Examples: Examples of discrete subgroups of $\mathrm{SL}_2(\mathbb{R})$ are provided by $\mathrm{SL}_2(\mathbb{Z})$ and each subgroup $\Gamma \subset \mathrm{SL}_2(\mathbb{Z})$. In particular, for $n \in \mathbb{N}$ we define the following subgroups

$$\Gamma_0(N) := \{A \in \mathrm{SL}_2(\mathbb{Z}) \mid A \equiv \begin{pmatrix} * & * \\ 0 & * \end{pmatrix} (N)\},$$

$$\Gamma^0(N) := \{A \in \mathrm{SL}_2(\mathbb{Z}) \mid A \equiv \begin{pmatrix} * & 0 \\ * & * \end{pmatrix} (N)\},$$

$$\Gamma_1(N) := \{A \in \mathrm{SL}_2(\mathbb{Z}) \mid A \equiv \begin{pmatrix} 1 & * \\ 0 & 1 \end{pmatrix} (N)\},$$

$$\Gamma(N) := \{A \in \mathrm{SL}_2(\mathbb{Z}) \mid A \equiv \begin{pmatrix} 1 & 0 \\ 0 & 1 \end{pmatrix} (N)\}.$$

The group $\Gamma(N)$ is called the principal congruence subgroup of level N. The following sequence is exact:

$$1 \to \Gamma(N) \hookrightarrow \mathrm{SL}_2(\mathbb{Z}) \to \mathrm{SL}_2(\mathbb{Z}/N\mathbb{Z}) \to 1,$$

$$A \mapsto A \bmod N.$$

Hence $\Gamma(N)$ is a normal subgroup in $\mathrm{SL}_2(\mathbb{Z})$, and has finite index in $\mathrm{SL}_2(\mathbb{Z})$ (so therefore do $\Gamma_0(N) \supset \Gamma_1(N) \supset \Gamma(N)$ etc.). In terms of the Euler φ-function

$$\varphi(N) := N \cdot \prod_{\substack{p \mid N \\ p \text{ prime}}} \left(1 - p^{-1}\right)$$

and

$$\psi(N) := N \cdot \prod_{\substack{p \mid N \\ p \text{ prime}}} \left(1 + p^{-1}\right)$$

we have more precisely:

$$[\mathrm{SL}_2(\mathbb{Z}) : \Gamma(N)] = N^3 \cdot \prod_{\substack{p \mid N \\ p \text{ prime}}} \left(1 - p^{-2}\right) = N \cdot \varphi(N) \cdot \psi(N), \tag{1}$$

$$[\mathrm{SL}_2(\mathbb{Z}) : \Gamma_1(N)] = N^2 \cdot \prod_{\substack{p \mid N \\ p \text{ prime}}} \left(1 - p^{-2}\right) = \varphi(N) \cdot \psi(N), \tag{2}$$

$$[\mathrm{SL}_2(\mathbb{Z}) : \Gamma_0(N)] = N \cdot \prod_{\substack{p \mid N \\ p \text{ prime}}} \left(1 + p^{-1}\right) = \psi(N). \tag{3}$$

For $k \in \mathbb{Z}$, $\mathrm{SL}_2(\mathbb{R})$ acts to the right on functions $f : \mathfrak{h} \to \mathbb{C}$: For $A = \left(\begin{smallmatrix} a & b \\ c & d \end{smallmatrix}\right) \in \mathrm{SL}_2(\mathbb{R})$ and $k \in \mathbb{Z}$ we define

$$(f|_k A)(\tau) := (c\tau + d)^{-k} \cdot f(A\tau).$$

That $(f, A) \mapsto f|_k A$ in fact defines an action follows from the following cocycle relation: With $AA' = A''$ we have

$$(cA'\tau + d) \cdot (c'\tau + d') = (c''\tau + d''),$$

where c, d, c', d' etc. denote the lower entries of A, A' etc.

Remark: $\mathrm{SL}_2(\mathbb{Z})$ acts transitively on $P_1(\mathbb{Q}) \subset P_1(\mathbb{C})$. For a subgroup $\Gamma \subset \mathrm{SL}_2(\mathbb{Z})$ of finite index, the orbits with respect to the action of Γ on $P_1(\mathbb{Q})$ are called the cusps of Γ, or the cusps of $\Gamma \backslash \mathfrak{h}$.

Definition: *Let $k \in \mathbb{Z}$ and $\Gamma \subset \mathrm{SL}_2(\mathbb{Z})$ be of finite index. A function $f : \mathfrak{h} \to \mathbb{C}$ is called a **modular form** of weight k on Γ if:*

(i)   *f is holomorphic on $\mathfrak{h}$,*
(ii) *$f|_k A = f$ for all $A \in \Gamma$,*
(iii) *for every $S \in \mathrm{SL}_2(\mathbb{Z})$, $f|_k S$ has a Fourier expansion of the form:*

$$(f|_k S)(\tau) = \sum_{n \geq 0} a_n \cdot e^{2\pi i \tau n / N} \tag{4}$$

for suitable a_n and suitable $N \in \mathbb{N}$.

Let $M_k(\Gamma)$ denote the complex vector space of all modular forms of weight k on Γ.

Remarks:

1) Let f be a function on $\mathfrak{h}$ satisfying (i) and (ii). Then, for each $S \in \mathrm{SL}_2(\mathbb{Z})$, $f|_k S$ is invariant under $S^{-1}\Gamma S$. Since $S^{-1}\Gamma S$ has finite index in $\mathrm{SL}_2(\mathbb{Z})$, there exists an $N \in \mathbb{Z}$, $N > 0$, such that $\left(\begin{smallmatrix} 1 & N \\ 0 & 1 \end{smallmatrix}\right)S^{-1}\Gamma S = S^{-1}\Gamma S$, i.e. $\left(\begin{smallmatrix} 1 & N \\ 0 & 1 \end{smallmatrix}\right) \in S^{-1}\Gamma S$. Therefore $(f|_k S)(\tau + N) = (f|_k S)(\tau)$, so there exists a holomorphic mapping $g : \{t \in \mathbb{C} \mid 0 < |t| < 1\} \to \mathbb{C}$ with the property $(f|_k S)(\tau) = g(e^{2\pi i \tau/N})$, i.e.

$$(f|_k S)(\tau) = \sum_{n=-\infty}^{\infty} a_n \cdot e^{2\pi i \tau n/N}$$

for suitable a_n. Thus we see that the restriction in (iii) of the definition of modular forms only consists of the requirement that $a_n = 0$ for $n < 0$, and not in the condition that a Fourier expansion exists. The expansion (4) is called a Fourier expansion at the cusp $\Gamma S\infty$.

2) Let f be a modular form of weight k on Γ, let $s \in P_1(\mathbb{Q})$, $s = S\infty$ for a suitable $S \in \mathrm{SL}_2(\mathbb{Z})$ and $f|_k S$ as in (4). Then one puts $\mathrm{ord}_s(f) := n_0/N$, where n_0 denotes the smallest integer with $a_{n_0} \neq 0$, and, for k even, one puts $f(s) := a_0$. The quantities $f(s)$ and $\mathrm{ord}_s(f)$ are independent of the special choice of S and only depend on s modulo Γ. For if $r \in \Gamma S\infty$, $R\infty = r$, then there exists a $G \in \Gamma$ with $G \cdot R\infty = S\infty$. Since now $S^{-1}GR$ leaves the point ∞ fixed, it is of the form $\left(\begin{smallmatrix} \pm 1 & * \\ 0 & \pm 1 \end{smallmatrix}\right)$ and so there are $\varepsilon \in \{\pm 1\}$ and $n_1 \in \mathbb{Z}$ with $G \cdot R = S \cdot \left(\begin{smallmatrix} \varepsilon & n_1 \\ 0 & \varepsilon \end{smallmatrix}\right)$. Therefore we have:

$$\begin{aligned}
(f|_k R)(\tau) = (f|_k GR)(\tau) &= \left(f|_k S \cdot \left(\begin{smallmatrix} \varepsilon & n_1 \\ 0 & \varepsilon \end{smallmatrix}\right)\right)(\tau) \\
&= \varepsilon^k \cdot (f|_k S)(\tau + \varepsilon n_1) \\
&= \varepsilon^k \cdot \sum_{n \geq 0} a_n \cdot e^{\varepsilon \cdot 2\pi i n n_1/N} \cdot e^{2\pi i n \tau/N}.
\end{aligned}$$

Modular forms f which vanish at all cusps, i.e. which satisfy $\mathrm{ord}_s(f) > 0$ for all $s \in P_1(\mathbb{Q})$, are called cusp forms. The space of cusp forms of weight k on Γ is denoted $S_k(\Gamma)$.

3) As mentioned above, the space $X := \Gamma \backslash \mathfrak{h}$ can be given a complex structure so that $\pi : \mathfrak{h} \to \Gamma \backslash \mathfrak{h}$ is holomorphic. By adjunction of the cusps $\Gamma \backslash P_1(\mathbb{Q})$ one can make X into a compact Riemann surface $\overline{X}$. For even k, the conditions (i) and (ii) correspond to the fact that there exists a holomorphic differential form ω of weight $k/2$ on X with $\pi^*\omega = f(\tau)(d\tau)^{k/2}$. Condition (iii) then implies that at the cusps ω has at most a pole, whose order is bounded by $k/2$. In particular, for $k = 2$ and f a cusp form, the differential form ω is holomorphic on all of $\overline{X}$, and vice versa any holomorphic differential on $\overline{X}$ can be so obtained.

Thus, one can interpret modular forms as sections of certain line bundles over $\overline{X}$; from general theorems on Riemann surfaces, the finite dimensionality of the spaces $M_k(\Gamma)$ then follows immediately (cf. [Sh71]). For $k = 2$, it follows from this point of view that the dimension of $S_2(\Gamma)$ is equal to the genus of X.

For general background on modular forms see, for example, [Sh71], [Se70], [Ap76], [Sch74] or [Gu62].

2 Examples of modular forms

2.1 Eisenstein series

Let $g : \mathfrak{h} \to \mathbb{C}$ be a mapping, then $f = \sum_{A \in \Gamma} g|_k A$ transforms formally as a modular form of weight k on Γ. It naturally remains to investigate the convergence of such an expression. A series defined in this manner is certainly divergent if the stabilizer $\Gamma_g = \{A \mid g|_k A = g\}$ of g has infinite order (and $g \neq 0$). A better definition for f is therefore in general:

$$f := \sum_{A \in \Gamma_g \backslash \Gamma} g|_k A.$$

For g a constant function and $\Gamma = \mathrm{SL}_2(\mathbb{Z})$ this construction yields, as an example, the so-called Eisenstein series:

$$E_k := \frac{1}{2} \cdot \sum_{A \in \left(\begin{smallmatrix} 1 & * \\ 0 & 1 \end{smallmatrix}\right) \backslash \mathrm{SL}_2(\mathbb{Z})} 1|_k A. \tag{1}$$

Note that for k odd $M_k(\mathrm{SL}_2(\mathbb{Z})) = 0$, since any $f \in M_k(\mathrm{SL}_2(\mathbb{Z}))$ satisfies $f = f|_k(-\mathrm{id}) = (-1)^k f$. The same argument holds of course for any Γ containing $-\mathrm{id}$. Thus we can restrict to even k, and here one has:

Theorem 2.1: *For $k > 2$ even, $E_k \in M_k(\mathrm{SL}_2(\mathbb{Z}))$. More precisely, the sum (1) is normally convergent on $\mathfrak{h}$, thus defines a holomorphic function on $\mathfrak{h}$, is invariant under $\mathrm{SL}_2(\mathbb{Z})$ with respect to the "$|_k$"-action, and one has*

$$E_k(\tau) = 1 + O\big(e^{2\pi i \tau}\big) \quad \text{for } \mathrm{Im}\,(\tau) \to \infty.$$

Proof: To show that the series in (1) is normally convergent, it suffices to show that it is absolutely and uniformly convergent in each region of the form B:

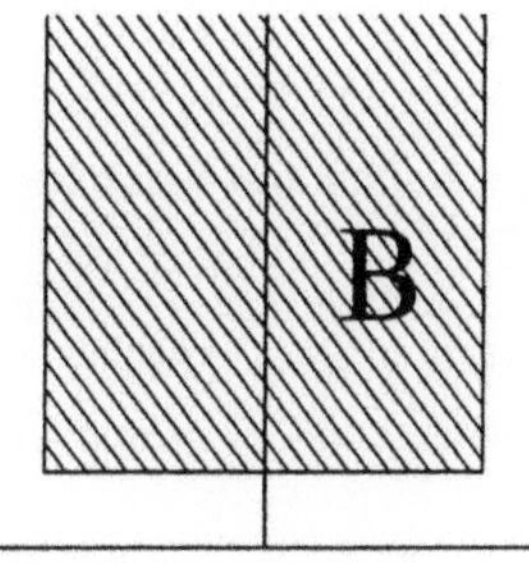

Region of the form B

From this it will also follow that :

$$\lim_{\nu \to \infty} E_k(i \cdot \nu) = \frac{1}{2} \sum_{A \in \left(\begin{smallmatrix} 1 & \ast \\ 0 & 1 \end{smallmatrix}\right) \backslash \mathrm{SL}_2(\mathbb{Z})} \lim_{\nu \to \infty} (1|_k A)(i \cdot \nu) = 1.$$

For the last identity note that $\lim_{\nu \to \infty} (1|_k A)(i \cdot \nu)$ is different from 0, and then equal to 1, if and only if $A \in \pm \left(\begin{smallmatrix} 1 & \mathbb{Z} \\ 0 & 1 \end{smallmatrix}\right)$. To prove the asserted convergence note first of all that two matrices A and A' lie in the same left coset with respect to $\left(\begin{smallmatrix} 1 & \mathbb{Z} \\ 0 & 1 \end{smallmatrix}\right)$ if and only if their lower rows are equal. Thus we can write

$$E_k(\tau) = \frac{1}{2} \sum_{\substack{c,d \in \mathbb{Z} \\ \gcd(c,d)=1}} \frac{1}{(c\tau + d)^k}. \tag{2}$$

Set

$$\mu := \inf_{\substack{x^2 + y^2 = 1 \\ \tau \in B}} |x\tau + y|^2.$$

Then μ is positive, since the continuous function $(\tau, x, y) \mapsto |x\tau + y|^2$ attains its minimum on the set of all (τ, x, y) with $\tau \in B$ and $x^2 + y^2 = 1$. Using this μ one has for all $c, d \in \mathbb{Z}$ and $\tau \in B$ the inequality

$$|c\tau + d|^2 \geq \mu(c^2 + d^2).$$

Thus the sum in (2) is dominated by

$$\mu^{-k/2} \sum_{\substack{c,d \in \mathbb{Z} \\ (c,d) \neq 0}} \frac{1}{(c^2 + d^2)^{k/2}},$$

which in turn is dominated by

$$\mathrm{const} \cdot \int_{\mathbb{R}^2 \backslash \{0\}} \frac{dx\,dy}{(x^2 + y^2)^{k/2}} = \mathrm{const} \cdot \int_0^{2\pi} \int_0^{\infty} \frac{r\,dr\,d\theta}{r^k} < \infty. \qquad \square$$

We have $M_{2k}(\mathrm{SL}_2(\mathbb{Z})) = \mathbb{C} \cdot E_{2k} \oplus S_{2k}(\mathrm{SL}_2(\mathbb{Z}))$. This follows easily from the facts that $E_{2k}(\infty) = 1$ and ∞ is the only cusp of $\mathrm{SL}_2(\mathbb{Z})$ (i.e. $\mathrm{SL}_2(\mathbb{Z})$ acts transitively on $P_1(\mathbb{Q})$).

2.2 Theta series

Definition: *Let k be even and let $L \subset \mathbb{R}^k$ be a lattice, i.e. there is a basis $(e_1, \ldots, e_k)$ of $\mathbb{R}^k$ so that $L = \mathbb{Z}e_1 + \cdots + \mathbb{Z}e_k$; assume that L is integral and even, i.e. $x \cdot y \in \mathbb{Z}$ and $x^2 \in 2\mathbb{Z}$ for all $x, y \in L$. Here $x \cdot y$ denotes the standard scalar product on $\mathbb{R}^k$. The **determinant** of L is defined by $\det(L) := \sharp\{L^{\mathrm{dual}}/L\}$, where $L^{\mathrm{dual}} = \{y \in \mathbb{R}^k \mid y \cdot L \subset \mathbb{Z}\}$. The **level** N of L is the smallest natural number N for which $Ny^2 \in 2\mathbb{Z}$ for all $y \in L^{\mathrm{dual}}$.*

Definition: *For a Dirichlet character χ modulo N we define:*

$$M_k(\Gamma_0(N), \chi) := \{f \in M_k(\Gamma_1(N)) \mid f|_k \left(\begin{smallmatrix} a & b \\ c & d \end{smallmatrix}\right) = \chi(d) \cdot f \text{ for all } \left(\begin{smallmatrix} a & b \\ c & d \end{smallmatrix}\right) \in \Gamma_0(N)\}.$$

Let

$$\theta_L(\tau) := \sum_{x \in L} e^{\pi i \tau x^2} = \sum_{n=0}^{\infty} \sharp\{x \in L \mid x^2/2 = n\} \cdot e^{2\pi i \tau n}.$$

Theorem 2.2: *We have*

$$\theta_L(\tau) \in M_{k/2}\left(\Gamma_0(N), \left(\frac{(-1)^{k/2} \det(L)}{\cdot}\right)\right),$$

where $\left(\frac{\cdot}{\cdot}\right)$ denotes the Jacobi symbol.

For the proof the reader is referred to [Sch74] or [Og69].

Example:

$$\theta(\tau) := \sum_{x,y \in \mathbb{Z}} e^{2\pi i \tau (x^2 + y^2)}$$

$$= \sum_{n=0}^{\infty} \sharp\{(x,y) \in \mathbb{Z}^2 \mid x^2 + y^2 = n\} \cdot e^{2\pi i \tau n}.$$

Here $\theta = \theta_L$ for $L = \sqrt{2} \cdot (\mathbb{Z} \times \mathbb{Z})$. We have $L^{\mathrm{dual}} = \frac{1}{\sqrt{2}}\mathbb{Z} \times \frac{1}{\sqrt{2}}\mathbb{Z}$, $N = 4$, $\det(L) = 4$, $k = 2$; therefore $\theta \in M_1\left(\Gamma_0(4), \left(\frac{-4}{\cdot}\right)\right)$. Moreover, from theorems of number theory one knows an explicit formula for the number of representations of an integer by the quadratic form $x^2 + y^2$. More precisely one has:

$$\theta(\tau) = 1 + 4 \cdot \sum_{n=1}^{\infty} \left(\sum_{d|n} \left(\tfrac{-4}{d}\right)\right) \cdot e^{2\pi i \tau n}.$$

For further information on theta series see [Sch74] and [Og69].

2.3 Lattice functions

Let $F : \{\text{lattices } L \subset \mathbb{C}\} \to \mathbb{C}$ be a homogeneous function of degree $-k$, i.e.

$$F(\lambda L) = F(L) \cdot \lambda^{-k} \quad \text{for all } \lambda \in \mathbb{C}^*.$$

An example for $k > 2$ is: $s_k(L) = \sum_{\gamma \in L \setminus \{0\}} \frac{1}{\gamma^k}$.

Now define $f(\tau) := F(\mathbb{Z}\tau + \mathbb{Z})$; then $f|_k A = f$ for all $A = \left(\begin{smallmatrix} a & b \\ c & d \end{smallmatrix}\right) \in \mathrm{SL}_2(\mathbb{Z})$, for we have:

$$(f|_k A)(\tau) = (c\tau + d)^{-k} \cdot f\left(\tfrac{a\tau + b}{c\tau + d}\right) = (c\tau + d)^{-k} \cdot F\left(\mathbb{Z}\,\tfrac{a\tau + b}{c\tau + d} + \mathbb{Z}\right)$$
$$= F(\mathbb{Z} \cdot (a\tau + b) + \mathbb{Z} \cdot (c\tau + d)) = F(\mathbb{Z}\tau + \mathbb{Z}) = f(\tau)$$

(note that with $(\tau, 1)$ also $(a\tau + b, c\tau + d)$ is a basis of $\mathbb{Z}\tau + \mathbb{Z}$).

On the other hand, if $f|_k A = f$ for all $A \in \mathrm{SL}_2(\mathbb{Z})$, then

$$F(\mathbb{Z}\omega_2 + \mathbb{Z}\omega_1) := \omega_1^{-k} \cdot f(\omega_2/\omega_1) \quad (\text{with } \mathrm{Im}\,(\omega_2/\omega_1) > 0)$$

defines a lattice function, homogeneous of degree $-k$. This connection can be generalized:

Lemma: *Let the function*

$$F : \{(L, \overline{\omega}_0) \mid L \text{ lattice}, \overline{\omega}_0 \text{ primitive } N\text{-division point of } \mathbb{C}/L\} \to \mathbb{C}$$

be homogeneous of degree $-k$, i.e.

$$F(\lambda L, \lambda \overline{\omega}_0) = \lambda^{-k} \cdot F(L, \overline{\omega}_0),$$

let f be defined by

$$f(\tau) := F\left(\mathbb{Z}\tau + \mathbb{Z}, \tfrac{1}{N}\right);$$

then $f|_k A = f$ for all $A \in \Gamma_1(N)$.

Proof: Here the second argument is always regarded as a class modulo L:

$$(f|_k A)(\tau) = F\left(\mathbb{Z}\,\tfrac{a\tau + b}{c\tau + d} + \mathbb{Z}, \tfrac{1}{N}\right) \cdot (c\tau + d)^{-k}$$
$$= F\left(\mathbb{Z}(a\tau + b) + \mathbb{Z}(c\tau + d), \tfrac{c\tau + d}{N}\right)$$
$$= F\left(\mathbb{Z}\tau + \mathbb{Z}, \tfrac{c}{N}\tau + \tfrac{d}{N}\right)$$
$$= F\left(\mathbb{Z}\tau + \mathbb{Z}, \tfrac{1}{N}\right) = f(\tau).$$

For the next to last equality observe that $c \equiv 0 \ (N)$ and $d \equiv 1 \ (N)$, hence $\frac{c}{N}\tau + \frac{d}{N} \equiv \frac{1}{N} \ (\mathbb{Z}\tau + \mathbb{Z})$.

On the other hand, if f transforms as a modular form on $\Gamma_1(N)$ of weight k, then one can define a function $F : \{(L, \overline{\omega}_0), \overline{\omega}_0 \text{ as above}\} \to \mathbb{C}$ by $F(L, \overline{\omega}_0) := \omega_1^{-k} \cdot f(\omega_2/\omega_1)$. Here $\omega_1 = N\omega_0$ and ω_2 is chosen so that $L = \mathbb{Z}\omega_2 + \mathbb{Z}\omega_1$ and $\mathrm{Im}\,(\omega_2/\omega_1) > 0$. That one can find such an ω_2 for a suitable choice of representative ω_0 of $\overline{\omega}_0$ is left to the reader as an exercise (or cf. Lemma 7.5).

Lemma: *This function F is well-defined and homogeneous of degree $-k$.*

3   The Weierstraß $\wp$-function as a Jacobi form

If one considers further the last two examples of the previous paragraph, one is led to
the theory of the so-called Jacobi forms (cf. [EiZa85]). We shall not define Jacobi forms
in complete generality, rather we shall consider only some special cases of interest to
us. We consider the following situation:

(J)   For $\tau \in \mathfrak{h}$, let $x \mapsto \Phi(\tau, x)$ be a meromorphic, periodic (hence elliptic) function
      with respect to the lattice $2\pi i(\mathbb{Z}\tau + \mathbb{Z})$. Further, let $k \in \mathbb{Z}$, $\Gamma \subset \mathrm{SL}_2(\mathbb{Z})$ of
      finite index, and assume that:

$$\Phi\left(A\tau, \tfrac{x}{c\tau+d}\right)(c\tau + d)^{-k} = \Phi(\tau, x) \quad \text{for } A = \left(\begin{smallmatrix} a & b \\ c & d \end{smallmatrix}\right) \in \Gamma.$$

Such functions on $\mathfrak{h} \times \mathbb{C}$ are called Jacobi forms on Γ of weight k (and index 0), if
certain additional regularity conditions are fulfilled.

Example: The Weierstraß $\wp$-function is defined as

$$\wp(\tau, x) := \frac{1}{x^2} + \sum_{\substack{\gamma \in 2\pi i(\mathbb{Z}\tau + \mathbb{Z}) \\ \gamma \neq 0}} \left(\frac{1}{(x - \gamma)^2} - \frac{1}{\gamma^2} \right),$$

and fulfills (J) with $\Gamma = \mathrm{SL}_2(\mathbb{Z})$ and $k = 2$.

Theorem 3.1: *Let $\Phi : \mathfrak{h} \times \mathbb{C} \to P_1(\mathbb{C})$ be a function which satisfies the assumptions* (J).
Let $n \in \mathbb{Z}$, $\alpha, \beta \in \mathbb{R}$ and let $g_n(\tau)$ be the n-th coefficient in the Taylor expansion of
$\Phi(\tau, \cdot)$ *at* $2\pi i(\alpha\tau + \beta)$. *Then we have:*

$$g_n|_{k+n} A = g_n \quad \text{for all } A \in \Gamma \text{ with } (\alpha, \beta)A \equiv (\alpha, \beta) \bmod \mathbb{Z}^2.$$

Proof: By the Cauchy integral formula,

$$g_n(\tau) = \frac{1}{2\pi i} \oint \frac{\Phi(\tau, x + 2\pi i(\alpha\tau + \beta))}{x^{n+1}}\, dx$$

and therefore

$$(g_n|_{k+n} A)(\tau) = g_n(A\tau)(c\tau + d)^{-(k+n)}$$

$$= \frac{1}{2\pi i} \oint \frac{\Phi(A\tau, (x(c\tau + d) + 2\pi i(\alpha'\tau + \beta'))/(c\tau + d))}{(x(c\tau + d))^{n+1}} (c\tau + d)^{-k}\, d(x(c\tau + d))$$

with $A = \left(\begin{smallmatrix} a & b \\ c & d \end{smallmatrix}\right)$ and $(\alpha', \beta') = (\alpha, \beta)A$. By substituting x for $x(c\tau + d)$ and using
(J), the assertion follows. $\square$

Example: Put $e_1(\tau) := \wp(\tau, \pi i)$. This therefore transforms as a modular form of
weight two with respect to $\{A \in \mathrm{SL}_2(\mathbb{Z}) \mid (0, \tfrac{1}{2})A \equiv (0, \tfrac{1}{2}) \bmod \mathbb{Z}^2\}$. The latter set
is nothing else than the group $\Gamma_0(2)$.

More generally, we want to consider the Taylor coefficients of $\wp(\tau, z)$ around any expansion point. For this, we prove the following

Theorem 3.2: *We have*

$$\wp(\tau, z) = \sum_{n \in \mathbb{Z}} \frac{1}{\left(q^{n/2}\zeta^{1/2} - q^{-n/2}\zeta^{-1/2}\right)^2} - \left(-\frac{1}{12} + \sum_{\substack{n \in \mathbb{Z} \\ n \neq 0}} \frac{1}{\left(q^{n/2} - q^{-n/2}\right)^2}\right). \qquad (1)$$

Here we make the convention, for now and the following, that:

$$q = e^{2\pi i \tau} \quad \text{and} \quad \zeta = e^{x}.$$

Proof: As is easily checked, the first summand is normally convergent, and, for fixed τ, it is periodic in x with respect to $2\pi i(\mathbb{Z}\tau + \mathbb{Z})$. It has poles precisely at the lattice points, each of order two; therefore for fixed τ the function defined by the first sum is $\wp(\tau, \cdot)$ up to addition of a constant. The asserted identity now follows from the facts that $\wp(x) = \frac{1}{x^2} + O(x)$ and that the function defined by the first sum is equal for x tending to 0 to

$$\frac{1}{\left(\zeta^{1/2} - \zeta^{-1/2}\right)^2} + \sum_{n \neq 0} \frac{1}{\left(q^{n/2} - q^{-n/2}\right)^2} + O(x)$$

$$= \frac{1}{x^2} - \frac{1}{12} + \sum_{n \neq 0} \frac{1}{\left(q^{n/2} - q^{-n/2}\right)^2} + O(x). \qquad \square$$

With the help of the identity

$$\frac{1}{\left(a^{1/2} - a^{-1/2}\right)^2} = \frac{a}{(1-a)^2} = \sum_{k=1}^{\infty} k \cdot a^k \qquad (|a| < 1)$$

we obtain immediately from (1) the

Lemma 3.3: *For* $|q| < \min\{|\zeta|, |\zeta^{-1}|\}$ *we have*

$$\wp(\tau, x) = \frac{1}{\left(\zeta^{1/2} - \zeta^{-1/2}\right)^2} + \sum_{n=1}^{\infty} \left(\sum_{d|n} d(\zeta^d + \zeta^{-d})\right) q^n + \frac{1}{12}\left(1 - 24\sum_{n=1}^{\infty} \sigma_1(n)q^n\right).$$

Here

$$\sigma_r(n) := \sum_{d|n} d^r. \qquad \square$$

At this point we shall introduce the Bernoulli numbers B_i, which are defined as follows:

$$\frac{x}{e^x - 1} = \sum_{i=0}^{\infty} \frac{B_i}{i!} x^i.$$

In particular, $B_0 = 1$, $B_1 = -\frac{1}{2}$ and $B_{2k+1} = 0$ for $k \geq 1$. Moreover, one has the following table:

k	1	2	3	4	5	6	7	8	9	10
B_{2k}	$\frac{1}{6}$	$\frac{-1}{30}$	$\frac{1}{42}$	$\frac{-1}{30}$	$\frac{5}{66}$	$\frac{-691}{2730}$	$\frac{7}{6}$	$\frac{-3617}{510}$	$\frac{43867}{798}$	$\frac{-174611}{330}$
$\frac{-4k}{B_{2k}}$	-24	240	-504	480	-264	$\frac{65520}{691}$	-24	$\frac{16320}{3617}$	$\frac{-28728}{43867}$	$\frac{13200}{174611}$

Bernoulli numbers

We further have

$$\frac{\coth(x/2)}{2} = \frac{1}{\zeta - 1} + \frac{1}{2} = \frac{1}{x} + \sum_{i=1}^{\infty} \frac{B_{2i}}{(2i)!} \, x^{2i-1}$$

and therefore

$$\frac{1}{\left(\zeta^{1/2} - \zeta^{-1/2}\right)^2} = \frac{1}{4\sinh(x/2)^2} = -\frac{d}{dx}\left(\frac{\coth(x/2)}{2}\right)$$

$$= \frac{1}{x^2} - \sum_{i=1}^{\infty} \frac{B_{2i}}{2i} \, \frac{x^{2i-2}}{(2i-2)!}. \tag{2}$$

By means of the Fourier expansion of the $\wp$-function we now obtain the

Theorem 3.4: *For $k > 1$,*

$$E_{2k} = 1 - \frac{4k}{B_{2k}} \cdot \sum_{n=1}^{\infty} \left(\sum_{d\mid n} d^{2k-1}\right) \cdot q^n$$

$$= 1 - \frac{4k}{B_{2k}} \cdot \sum_{n=1}^{\infty} \sigma_{2k-1}(n) \cdot q^n.$$

Proof: Consider the Laurent expansion of $\wp(\tau, x)$ around $x = 0$. On the one hand, directly from the definition of the $\wp$-function and the Taylor formula, we find the

$(2k-2)$-nd coefficient to be

$$\frac{(2k-1)!}{(2k-2)!} \cdot \sum_{\substack{\gamma \in 2\pi i(\mathbb{Z}\tau+\mathbb{Z}) \\ \gamma \neq 0}} \frac{1}{\gamma^{2k}} = \frac{2k-1}{(2\pi i)^{2k}} \cdot \sum_{\substack{m,n \in \mathbb{Z} \\ (m,n) \neq 0}} \frac{1}{(m\tau+n)^{2k}}$$

$$= \frac{2k-1}{(2\pi i)^{2k}} \cdot \sum_{n=1}^{\infty} \frac{1}{n^{2k}} \sum_{\substack{c,d \in \mathbb{Z} \\ (c,d) \neq 0 \\ \gcd(c,d)=1}} \frac{1}{(c\tau+d)^{2k}}$$

$$= \frac{2k-1}{(2\pi i)^{2k}} \cdot \zeta(2k) \cdot 2E_{2k}(\tau),$$

the latter directly from the definition of E_{2k} (cf. §2, formula (2)) and the definition of the Riemann ζ-function $\zeta(s) := \sum_{n \geq 1} n^{-s}$.

On the other hand, it follows from Lemma 3.3 and (2) that the $(2k-2)$-nd coefficient of the $\wp$-function is none other than

$$-\frac{2k-1}{(2k)!} B_{2k} + \frac{2}{(2k-2)!} \cdot \sum_{n=1}^{\infty} \sigma_{2k-1}(n) \cdot q^n.$$

A comparison of the two expressions yields

$$\frac{2k-1}{(2\pi i)^{2k}} \cdot \zeta(2k) \cdot 2E_{2k} = -\frac{2k-1}{(2k)!} B_{2k} + \frac{2}{(2k-2)!} \cdot \sum_{n=1}^{\infty} \sigma_{2k-1}(n) \cdot q^n.$$

Since $E_{2k} = 1 + O(q)$ by Theorem 2.1, the assertion follows at once, as does the following corollary. $\qquad\qquad\square$

Corollary 3.5 (Euler): *For the Riemann ζ-function, we have*

$$\zeta(2k) = -\frac{1}{2} \frac{(2\pi i)^{2k}}{(2k)!} B_{2k}. \qquad\qquad\square$$

Remark: This also holds for $k = 1$, on the basis of other considerations.

We now define

$$E_2(\tau) := 1 - 24 \sum_{n=1}^{\infty} \sigma_1(n) \cdot q^n$$

and

$$G_{2k}(\tau) := -\frac{B_{2k}}{4k} E_{2k} = -\frac{B_{2k}}{4k} + \sum_{n=1}^{\infty} \sigma_{2k-1}(n) \cdot q^n, \tag{3}$$

or equivalently (in view of Theorem 3.4 and its corollary)

$$G_{2k}(\tau) := \frac{\zeta(2k)}{(2\pi i)^{2k}} (2k-1)! \cdot E_{2k},$$

hence for $k \geq 2$

$$G_{2k}(\tau) = \frac{(2k-1)!}{2} \cdot \sum_{\substack{\gamma \in 2\pi i(\mathbb{Z}\tau + \mathbb{Z}) \\ \gamma \neq 0}} \frac{1}{\gamma^{2k}} . \tag{4}$$

We thereby obtain as an alternative formulation of Lemma 3.3:

$$\wp(\tau, x) = \frac{1}{x^2} + 2 \sum_{k=2}^{\infty} G_{2k}(\tau) \frac{x^{2k-2}}{(2k-2)!} .$$

As a further consequence of Lemma 3.3 and as an illustration of Theorem 3.1 one has

Theorem 3.6: *The constant terms* $e_1(\tau) := \wp(\tau, \pi i)$, $e_2(\tau) := \wp(\tau, \pi i\tau)$, *and* $e_3(\tau) := \wp(\pi i(\tau + 1))$ *of the Taylor expansion at the primitive two-division points are modular forms of weight* 2 *(with respect to different groups). They are permuted by the* "$\vert_2$"*-operation of* $\mathrm{SL}_2(\mathbb{Z})$, $e_1(\tau)$ *being invariant under* $\Gamma_0(2)$. *We have:*

$$e_1(\tau) = -\frac{1}{6} \cdot \Big(1 + 24 \cdot \sum_{n=1}^{\infty} \sigma_1^{\mathrm{odd}}(n)q^n\Big) \in M_2(\Gamma_0(2)),$$

$$e_2(\tau) = \frac{1}{12} \cdot \Big(1 + 24 \cdot \sum_{n=1}^{\infty} \sigma_1^{\mathrm{odd}}(n)q^{n/2}\Big) \in M_2(\Gamma^0(2)),$$

$$e_3(\tau) = \frac{1}{12} \cdot \Big(1 + 24 \cdot \sum_{n=1}^{\infty} \sigma_1^{\mathrm{odd}}(n)(-1)^n q^{n/2}\Big) \in M_2\big(\big(\begin{smallmatrix} 1 & 1 \\ -1 & 0 \end{smallmatrix}\big)\Gamma_0(2)\big(\begin{smallmatrix} 0 & -1 \\ 1 & 1 \end{smallmatrix}\big)\big).$$

Here $\sigma_1^{\mathrm{odd}}(n) = \sum_{\substack{d \mid n \\ d \equiv 1\ (2)}} d.$

Proof: That e_1, e_2, and e_3 transform as modular forms for $\Gamma_0(2)$, $\Gamma^0(2)$, etc., follows from Theorem 3.1. That they are holomorphic on $\mathfrak{h}$ and at the cusp ∞ can be derived from their Fourier expansions, which in turn are given by Lemma 3.3. That they are permuted by $\mathrm{SL}_2(\mathbb{Z})$ follows from $\wp\big(A\tau, \frac{x}{c\tau+d}\big)(c\tau + d)^{-2} = \wp(\tau, x)$, for $A = \big(\begin{smallmatrix} a & b \\ c & d \end{smallmatrix}\big) \in \mathrm{SL}_2(\mathbb{Z})$, whence for $\wp_{(\alpha,\beta)}(\tau) := \wp(\tau, \alpha\tau + \beta)$ we have

$$\wp_{(\alpha,\beta)}\vert_2(A)(\tau) := \wp(A\tau, \alpha A\tau + \beta)(c\tau + d)^{-2}$$
$$= \wp(\tau, \alpha'\tau + \beta')$$
$$= \wp_{(\alpha',\beta')}$$

for $(\alpha', \beta') = (\alpha, \beta)A$. The regularity at all other cusps also follows from this property of being permuted. $\qquad \Box$

Remark 3.7: In particular we have, as will be frequently useful for the study of elliptic genera:

$$\delta := -\frac{3}{2} e_1 = \frac{1}{4} + 6 \cdot \sum_{n=1}^{\infty} \sigma_1^{\mathrm{odd}}(n) q^n$$

$$= \frac{1}{4} + 6 \cdot \left(q + q^2 + 4q^3 + q^4 + 6q^5 + 4q^6 + \cdots \right) \in M_2(\Gamma_0(2)),$$

$$\widetilde{\delta} := \frac{3}{4} e_1 = -\frac{1}{8} - 3 \cdot \sum_{n=1}^{\infty} \sigma_1^{\mathrm{odd}}(n) q^n \in M_2(\Gamma_0(2)),$$

$$\varepsilon := (e_1 - e_2)(e_1 - e_3)$$

$$= \frac{1}{16} - q + 7q^2 - 28q^3 + 71q^4 - 126q^5 + 196q^6 \pm \cdots \in M_4(\Gamma_0(2)),$$

$$\widetilde{\varepsilon} := \frac{1}{16} (e_2 - e_3)^2$$

$$= q + 8q^2 + 28q^3 + 64q^4 + 126q^5 + 224q^6 + \cdots \in M_4(\Gamma_0(2)),$$

$$\iota := (e_1 - e_2)(e_1 - e_3)(e_2 - e_3)^2$$

$$= q - 8q^2 + 12q^3 + 64q^4 - 210q^5 \pm \cdots \in S_8(\Gamma_0(2)),$$

$$\Delta := 16 \cdot \prod_{i<j} (e_i - e_j)^2$$

$$= q - 24q^2 + 252q^3 - 1472q^4 + 4830q^5 - 6048q^6 \pm \cdots \in S_{12}(\mathrm{SL}_2(\mathbb{Z})).$$

In each case, the property of being a modular form or cusp form for $\Gamma_0(2)$ follows from the permutation statement of Theorem 3.6.

One can immediately derive the following relations, also useful for the study of elliptic genera:

$$e_1 + e_2 + e_3 = 0, \tag{5}$$

$$\widetilde{\delta} = -\frac{1}{2}\,\delta, \quad \delta = -2\widetilde{\delta}, \tag{6}$$

$$\widetilde{\varepsilon} = \frac{1}{4} \left(\delta^2 - \varepsilon \right), \quad \varepsilon = 4\left(\widetilde{\delta}^2 - \widetilde{\varepsilon} \right), \tag{7}$$

$$\iota = 16\varepsilon\widetilde{\varepsilon} = 4\varepsilon\left(\delta^2 - \varepsilon \right) = 64\widetilde{\varepsilon}\left(\widetilde{\delta}^2 - \widetilde{\varepsilon} \right), \tag{8}$$

$$\Delta = 256\varepsilon^2\widetilde{\varepsilon} = 16\varepsilon\iota = 64\varepsilon^2\left(\delta^2 - \varepsilon \right) = 4096\widetilde{\varepsilon}\left(\widetilde{\delta}^2 - \widetilde{\varepsilon} \right)^2, \tag{9}$$

$$e_1(\tau) = -2e_2(2\tau). \tag{10}$$

Formula (5) follows from the Fourier expansions of e_1, e_2 and e_3; (6) directly from the definition; (7) directly from the definition and (5); (8) and (9) directly from the definitions, resp. (7); (10) is clear.

4 Some special functions and modular forms

4.1 The valence formula

For the derivation of the explicit description of the modular forms on $\Gamma_0(2)$ given below, we first need a formula for $\dim M_k(\Gamma_0(2))$. One obtains this by means of the valence formula, valid for arbitrary subgroups $\Gamma \subset \mathrm{SL}_2(\mathbb{Z})$ of finite index.

Theorem 4.1 (valence formula): *Let $f \in M_k(\Gamma)$, $f \neq 0$, then*

$$\sum_{s \in \Gamma \backslash P_1(\mathbb{Q})} [\mathrm{PSL}_2(\mathbb{Z})_s : \mathrm{P}\Gamma_s] \cdot \mathrm{ord}_s(f) + \sum_{\tau \in \Gamma \backslash \mathfrak{h}} \frac{1}{|\mathrm{P}\Gamma_\tau|} \, \mathrm{ord}_\tau(f) = \frac{k}{12} [\mathrm{PSL}_2(\mathbb{Z}) : \mathrm{P}\Gamma].$$

In the first two sums, one is to add over a complete system of representatives for the orbits of $P_1(\mathbb{Q})$, resp. $\mathfrak{h}$, modulo Γ. Further, $\mathrm{PSL}_2(\mathbb{Z})$, $\mathrm{P}\Gamma$, etc. denotes the image of $\mathrm{SL}_2(\mathbb{Z})$, Γ, etc. under the natural map $\mathrm{SL}_2(\mathbb{R}) \to \mathrm{SL}_2(\mathbb{R})/\{\pm 1\}$, and $\mathrm{P}\Gamma_s$, $\mathrm{P}\Gamma_\tau$ are the subgroups of those elements in $\mathrm{P}\Gamma$ which leave s and τ invariant.

Remarks:

(i)   The first sum is finite:

$$\sharp(\Gamma \backslash P_1(\mathbb{Q})) \leq \sharp(\mathrm{SL}_2(\mathbb{Z}) \backslash P_1(\mathbb{Q})) \cdot [\mathrm{PSL}_2(\mathbb{Z}) : \mathrm{P}\Gamma]$$
$$= [\mathrm{PSL}_2(\mathbb{Z}) : \mathrm{P}\Gamma] < \infty.$$

(ii) In the second sum, almost all terms are equal to zero (compare the proof-sketch below).

(iii) The index $[\mathrm{PSL}_2(\mathbb{Z})_s : \mathrm{P}\Gamma_s]$ depends only on the class Γs. It is finite as is easily deduced from the fact that $[\mathrm{PSL}_2(\mathbb{Z}) : \mathrm{P}\Gamma] < \infty$.

(iv) We have $|\mathrm{P}\Gamma_\tau| < \infty$, for as one can easily prove (cf. Lemma 7.8), we have (with $\rho = e^{2\pi i/3}$):
If $\mathrm{P}\Gamma_\tau \neq 1$, then either

$$\tau \equiv i \bmod \mathrm{SL}_2(\mathbb{Z}) \text{ or}$$
$$\tau \equiv \rho \bmod \mathrm{SL}_2(\mathbb{Z}).$$

Further, $\mathrm{PSL}_2(\mathbb{Z})_i = \langle \left(\begin{smallmatrix} 0 & -1 \\ 1 & 0 \end{smallmatrix}\right) \rangle$ (with order 2) and $\mathrm{PSL}_2(\mathbb{Z})_\rho = \langle \left(\begin{smallmatrix} 0 & -1 \\ 1 & 1 \end{smallmatrix}\right) \rangle$ (with order 3).

Sketch of proof: The proof of the valence formula is essentially obtained by an application of the formula of Rouché:

$$N - P = \frac{1}{2\pi i} \cdot \int\limits_{\partial G} d\log(f),$$

where N is the number of zeroes and P is the number of poles of the meromorphic function f in the region G. As region G one takes a fundamental domain for $\mathfrak{h}$

modulo Γ (the analog of the fundamental mesh of a lattice), from which one has removed neighborhoods of the cusps which are so small that they contain no zeroes and poles. This is possible since f has a Fourier expansion at each cusp. Moreover, one has to take care of possible zeroes of f on the boundary of G. To compute the integral of $d\log(f)$ along ∂G one chooses a fundamental domain whose boundary consists of finitely many pairs of curves so that for each such pair there exists a transformation in Γ identifying the curves of this pair with reversed orientation. The integral of $d\log(f)$ along each such pair is then easily computed by using the transformation law of f under Γ.

For the sketched computation one may restrict to the case of a modular form on $\mathrm{SL}_2(\mathbb{Z})$. This is due to the following observations: Let k be even. Then

$$g := \prod_{A \in \mathrm{P}\Gamma \backslash \mathrm{PSL}_2(\mathbb{Z})} f|_k A$$

(A running through a complete set of representations for the $\mathrm{P}\Gamma$-left cosets in $\mathrm{PSL}_2(\mathbb{Z})$) defines a modular form of weight $k \cdot [\mathrm{PSL}_2(\mathbb{Z}) : \mathrm{P}\Gamma]$ on the full modular group $\mathrm{SL}_2(\mathbb{Z})$. Hence the right hand side of the valence formula for f on Γ equals the right hand side of the valence formula for g, considered as a modular form on $\mathrm{SL}_2(\mathbb{Z})$, and, as can easily be checked, the same holds true for the corresponding left hand sides. Thus the valence formula for g implies the one for f. This argument can be modified to include the case of odd k. For a complete proof of the valence formula for $\Gamma = \mathrm{SL}_2(\mathbb{Z})$ cf. [Se70], or, for the general case, cf. [Ra77], Theorem 4.14. ⌑

As a first consequence of the valence formula, we have the following

Corollary 4.2:

1) $M_k(\Gamma) = 0$ *for all* $k < 0$.
2) $\dim M_k(\Gamma) \leq 1 + \frac{k}{12} [\mathrm{PSL}_2(\mathbb{Z}) : \mathrm{P}\Gamma]$ *for* $k \geq 0$, *in particular* $M_0(\Gamma) = \mathbb{C}$.

Proof: Part 1) is a trivial consequence of the valence formula since its left hand side is always non-negative. To prove 2), note that for $\tau_1, \ldots, \tau_N \in \mathfrak{h}$ pairwise inequivalent modulo Γ, each having trivial stabilizer in $\mathrm{P}\Gamma$, the homomorphism

$$M_k(\Gamma) \to \mathbb{C}^N$$
$$f \mapsto (f(\tau_1), \ldots, f(\tau_N))$$

is injective, as soon as N is greater than the right side of the valence formula. Indeed, if $f \neq 0$ then the left hand side of the valence formula for f is greater or equal to the number of i's with $f(\tau_i) = 0$, on the other hand it is strictly less than N, whence f cannot vanish for all the τ_i. ⌑

4.2 The ring $M_*(2)$

If f and g are modular forms on a group Γ of weight k and l, respectively, then $f \cdot g$ is a modular form of weight $k+l$ on Γ. Hence, the subspace $M_*(\Gamma)$ of the space of all functions on $\mathfrak{h}$ which is spanned by all modular forms of all weights on Γ is a ring. Moreover, as can be easily checked, it is a graded ring, i.e.

$$M_*(\Gamma) = \bigoplus_{k \in \mathbb{Z}} M_k(\Gamma).$$

In this section we determine the structure of this ring in the case $\Gamma = \Gamma_0(2)$. We write $M_*(2)$ for $M_*(\Gamma_0(2))$. For $\Gamma = \Gamma_0(2)$, the space $\Gamma \backslash \mathfrak{h}$ has exactly two cusps (i.e. $\Gamma \backslash P_1(\mathbb{Q})$ has two elements), having representatives 0 and ∞ (cf. Lemma 7.11 below). Each $\tau \in \mathfrak{h}$ with $\mathrm{P}\Gamma_\tau \neq 1$ is equivalent modulo Γ to $e := (1+i)/2$ with $\Gamma_e = \langle \left(\begin{smallmatrix} 1 & -1 \\ 2 & -1 \end{smallmatrix} \right) \rangle$ (cf. Lemma 7.10 below). Finally, we have $[\mathrm{PSL}_2(\mathbb{Z}) : \mathrm{P}\Gamma_0(2)] = 3$ (cf. §1, formula (3)). Therefore the valence formula for $\Gamma = \Gamma_0(2)$ can be written as:

$$2 \operatorname{ord}_0(f) + \operatorname{ord}_\infty(f) + \frac{1}{2} \operatorname{ord}_e(f) + \sum_{\substack{\tau \in \Gamma \backslash \mathfrak{h} \\ \tau \not\equiv e \bmod \Gamma}} \operatorname{ord}_\tau(f) = \frac{k}{4}.$$

Using the special modular forms on $\Gamma = \Gamma_0(2)$ introduced in Remark 3.7, we now conclude:

1) $M_0(2) = \mathbb{C}$, $M_2(2) = \mathbb{C} \cdot \delta$, $M_4(2) = \mathbb{C} \cdot \delta^2 \oplus \mathbb{C} \cdot \varepsilon$ as a consequence of Corollary 4.2. From the Fourier expansions it follows that $\varepsilon(\infty)$, $\delta(\infty) \neq 0$ and $\operatorname{ord}_0(\varepsilon) = 1/2$ (cf. Theorem 3.6 and Remark 3.7). From this and the valence formula one deduces that $\delta(e) = 0$ but $\varepsilon(e) \neq 0$.

2) $M_k(2) \overset{\cdot \iota}{\to} S_{k+8}(2)$ is an isomorphism. The injectivity is clear. The surjectivity follows from the fact that by Remark 3.7 ι vanishes at the cusps, and has there the minimal vanishing orders for Γ, namely $\operatorname{ord}_0(\iota) = 1/2$, $\operatorname{ord}_\infty(\iota) = 1$, and that it has, by the valence formula, no further zeroes, whence any cusp form on $\Gamma_0(2)$ is divisible by ι (in the ring $M_*(2)$).

3) δ and ε are algebraically independent: Since $M_*(2)$ is a graded ring and δ and ε are homogeneous elements (of degree 2 and 4, respectively) it suffices to show that there is no non-trivial weighted homogeneous polynomial P such that $P(\delta, \varepsilon) = 0$. Assume that P is such a polynomial, say of minimal degree $2n$. Then write $P(\delta, \varepsilon) = a \cdot \varepsilon^{n/2} + Q(\delta, \varepsilon) \cdot \varepsilon \delta + b \cdot \delta^n$. Evaluation of P at the point e gives $a = 0$; at the zero of ε (the cusp $\Gamma 0$) it gives $b = 0$. Therefore, $Q(\delta, \varepsilon) = 0$, and Q has degree less than P. Contradiction!

We can now prove the following structure theorem:

Theorem 4.3: $M_*(2) = \mathbb{C}[\delta, \varepsilon]$.

Proof: By 3), $\mathbb{C}[\delta, \varepsilon] \subset M_*(2)$ is a graded polynomial ring. The dimension of the component of degree k in $\mathbb{C}[\delta, \varepsilon]$ is $1 + [k/4]$ for even k (0 otherwise). But Corollary 4.2 implies $\dim M_k(\Gamma) \leq 1 + k/4$.   ⌧

By 2), there is a practical algorithm for writing each $f \in M_k(2)$ as a polynomial $f = P(\delta, \varepsilon)$. Namely, $\iota = 4 \cdot (\delta^2 - \varepsilon)\varepsilon$ and to each $f \in M_k(\Gamma_0(2))$ one can find by consideration of the values at the cusps a simple polynomial $Q(\delta, \varepsilon)$ so that $f - Q(\delta, \varepsilon)$ is a cusp form (for the latter, note that: $\mathrm{ord}_0(\delta) = \mathrm{ord}_\infty(\delta) = 0$ and $\mathrm{ord}_0(\varepsilon) = 1/2$, $\mathrm{ord}_\infty(\varepsilon) = 0$).

Lemma 4.4: *Let* Γ, $\widetilde{\Gamma}$ *be subgroups of* $\mathrm{SL}_2(\mathbb{Z})$ *with* $\widetilde{\Gamma} \subset T^{-1}\Gamma T$ *for a matrix* $T \in \mathrm{SL}_2(\mathbb{R})$. *Further, let* $f(\tau)$ *be a modular form of weight* k *on* Γ. *Then* $f|_k T$ *is a modular form on* $\widetilde{\Gamma}$.

Proof: Let $\widetilde{A} \in \widetilde{\Gamma}$, $\widetilde{A} = T^{-1}AT$, $A \in \Gamma$. For $\widetilde{f} := f|_k T$ we then have:

$$\widetilde{f}|_k \widetilde{A} = \widetilde{f}|_k T^{-1}AT = f|_k TT^{-1}AT = f|_k AT$$
$$= f|_k T = \widetilde{f}.$$

The regularity at the cusps follows similarly, and the holomorphicity of $\widetilde{f}$ is clear. ⌧

In particular, by this lemma the normalizer of Γ in $\mathrm{SL}_2(\mathbb{R})$ acts on $M_k(\Gamma)$. For $\Gamma = \Gamma_0(2)$ the following special cases are of interest to us.

Corollary 4.5:

1) *For* $f(\tau) \in M_k(\mathrm{SL}_2(\mathbb{Z}))$ *(in particular,* $f \in M_k(\Gamma^0(2))$ *), we have* $f(2\tau) \in M_k(\Gamma_0(2))$.

2) *An involution* w_2 *is defined on* $M_k(\Gamma_0(2))$ *by* $f|w_2 := f|_k \left(\begin{smallmatrix} 0 & -1/\sqrt{2} \\ \sqrt{2} & 0 \end{smallmatrix}\right)$, *the so-called Fricke-Atkin-Lehner involution.* ⌧

In the main part of the book, for a given modular form $f \in M_k(\Gamma_0(2))$ one occasionally considers the special expansion

$$f^0 := f|_k \left(\begin{smallmatrix} 0 & -1 \\ 1 & 0 \end{smallmatrix}\right)$$

of f at the cusp $\Gamma_0(2)0$. Obviously we have

$$f^0(2\tau) = \frac{1}{2^{k/2}} f^0|_k \left(\begin{smallmatrix} \sqrt{2} & 0 \\ 0 & 1/\sqrt{2} \end{smallmatrix}\right)(\tau) = 2^{-k/2} \cdot (f|w_2)(\tau). \tag{1}$$

Theorem 4.6: *For the special expansions* e_1^0, δ^0 *and* ε^0 *of* e_1, δ *and* ε *at the cusp* $\Gamma_0(2)0$ *we have:*

$$e_1^0(2\tau) = -\frac{1}{2} e_1(\tau), \tag{2}$$

$$\delta^0(2\tau) = -\frac{1}{2} \delta(\tau) = \widetilde{\delta}(\tau), \tag{3}$$

$$\varepsilon^0(2\tau) = \widetilde{e}(\tau). \tag{4}$$

Proof: The identities for e_1 and δ follow from (1), since w_2 acts as $\pm$ id on the one-dimensional space M_2. The sign can be determined by taking for τ that element of $\mathfrak{h}$ satisfying $-1/\tau = 2\tau$. Now (2) and (3) follow directly. From Theorem 3.6 we know that the passage from τ to $-1/\tau$ permutes the e_i; more precisely, e_1 and e_2 are exchanged while e_3 is left fixed. Therefore we obtain:

$$\varepsilon|_4 \begin{pmatrix} 0 & -1 \\ 1 & 0 \end{pmatrix} = (e_2 - e_1)(e_2 - e_3) = q^{1/2} + 8q + \cdots,$$

from which (4) follows. $\qquad\qquad\qquad\qquad\qquad\qquad\qquad\qquad\qquad\qquad\qquad$ ⌸

Theorem 4.7: *The modular forms ε and $\widetilde{\varepsilon}$ in $M_4(\Gamma_0(2))$ have the following Fourier expansions:*

$$\varepsilon = \frac{1}{16} + \sum_{n \geq 1} \sum_{d \mid n} (-1)^d d^3 \cdot q^n, \tag{5}$$

$$\widetilde{\varepsilon} = \sum_{n \geq 1} \sum_{\substack{d \mid n \\ n/d \equiv 1\,(2)}} d^3 \cdot q^n. \tag{6}$$

Proof: By Lemma 4.4, and since $\dim M_4(\Gamma_0(2)) = 2$, $E_4(\tau)$ and $E_4(2\tau)$ generate the space $M_4(\Gamma_0(2))$. We have

$$E_4(2\tau) = 1 + 240 \sum_{n \geq 1} \sigma_3(n) \cdot q^{2n} \tag{7}$$

$$= 1 + 240 \sum_{n \geq 1} \sum_{\substack{d \mid n \\ n/d \equiv 0\,(2)}} d^3 \cdot q^n \tag{8}$$

$$= 1 + 30 \sum_{n \geq 1} \sum_{\substack{d \mid n \\ d \equiv 0\,(2)}} d^3 \cdot q^n. \tag{9}$$

From $\varepsilon = \frac{1}{16} - q + \cdots$ and (9) follows $\varepsilon = -\frac{1}{240} E_4(\tau) + \frac{2}{30} E_4(2\tau)$, and therefore (5); from $\widetilde{\varepsilon} = 0 + q + \cdots$ and (7) follows $\widetilde{\varepsilon} = \frac{1}{240} E_4(\tau) - \frac{1}{240} E_4(2\tau)$, and hence (6). ⌸

4.3 G_2 and the Dedekind η-function

Lemma 4.8: *The Eisenstein series*

$$G_2 = -\frac{1}{24} + \sum_{n=1}^{\infty} \sigma_1(n) q^n$$

transforms as follows (with $G_2^*(\tau) := G_2(\tau) + \frac{1}{8\pi \cdot \operatorname{Im}(\tau)}$):

$$G_2|_2 \begin{pmatrix} 0 & -1 \\ 1 & 0 \end{pmatrix}(\tau) = G_2(\tau) - \frac{1}{4\pi i \tau} \tag{10}$$

resp.

$$G_2^*|_2 A = G_2^* \tag{11}$$

for all $A \in \mathrm{SL}_2(\mathbb{Z})$.

Proof (after Hurwitz): Since $\mathrm{SL}_2(\mathbb{Z})$ is generated by $\begin{pmatrix} 0 & -1 \\ 1 & 0 \end{pmatrix}$ and $\begin{pmatrix} 1 & 1 \\ 0 & 1 \end{pmatrix}$, and the transformation behavior of G_2^* under the latter matrix is clear, the general assertion (11) follows easily from (10). For the proof of (10) we consider the function $G(\tau) := -4\pi^2 G_2(\tau)$. From the partial fraction expansion of $\pi^2/\sin^2$ follows:

$$
\begin{aligned}
G(\tau) &= -4\pi^2 \cdot \left(-\frac{1}{24} + \sum_{n=1}^{\infty} \frac{1}{\left(q^{n/2} - q^{-n/2} \right)^2} \right) \\
&= \frac{\pi^2}{6} + \sum_{n=1}^{\infty} \frac{\pi^2}{\sin\left(\pi n \tau \right)^2} \\
&= \frac{\pi^2}{6} + \sum_{n=1}^{\infty} \left(\frac{1}{n^2 \tau^2} + \sum_{k=1}^{\infty} \left(\frac{1}{(n\tau + k)^2} + \frac{1}{(n\tau - k)^2} \right) \right) \\
&= \frac{\pi^2}{6} + \frac{\pi^2}{6\tau^2} + \sum_{n=1}^{\infty} \sum_{k=1}^{\infty} \left(\frac{1}{(n\tau + k)^2} + \frac{1}{(n\tau - k)^2} \right),
\end{aligned}
$$

where we have used $\xi(2) = \pi^2/6$. Consequently,

$$\frac{1}{\tau^2} G(-1/\tau) = \frac{\pi^2}{6\tau^2} + \frac{\pi^2}{6} + \sum_{k=1}^{\infty} \sum_{n=1}^{\infty} \left(\frac{1}{(n\tau - k)^2} + \frac{1}{(n\tau + k)^2} \right).$$

Putting $g_{n,k} = \frac{1}{(n\tau + k)^2}$, the difference is therefore

$$G(\tau) - \tau^{-2} G(-1/\tau) = \sum_{n=1}^{\infty} \sum_{k=1}^{\infty} (g_{n,k} + g_{n,-k}) - \sum_{k=1}^{\infty} \sum_{n=1}^{\infty} (g_{n,k} + g_{n,-k}).$$

For positive k, define

$$h_{n,k} := \frac{1}{(n\tau + k - 1)(n\tau + k)} = \frac{1}{n\tau + k - 1} - \frac{1}{n\tau + k},$$

and

$$h_{n,-k} := \frac{1}{(n\tau - k)(n\tau - k + 1)} = \frac{-1}{n\tau - k + 1} + \frac{1}{n\tau - k};$$

then

$$g_{n,k} - h_{n,k} = \frac{\pm 1}{(n\tau + k)^2 (n\tau + k \pm 1)}$$

is absolutely summable. Therefore the following exchange of order of summation is permitted:

$$\sum_{n=1}^{\infty}\sum_{k=1}^{\infty}(g_{n,k}+g_{n,-k}) - \sum_{n=1}^{\infty}\sum_{k=1}^{\infty}(h_{n,k}+h_{n,-k})$$

$$= \sum_{k=1}^{\infty}\sum_{n=1}^{\infty}(g_{n,k}+g_{n,-k}) - \sum_{k=1}^{\infty}\sum_{n=1}^{\infty}(h_{n,k}+h_{n,-k}),$$

and the difference we are attempted to evaluate is:

$$\sum_{n=1}^{\infty}\sum_{k=1}^{\infty}(h_{n,k}+h_{n,-k}) - \sum_{k=1}^{\infty}\sum_{n=1}^{\infty}(h_{n,k}+h_{n,-k}).$$

Now the first sum is 0, since one verifies immediately that the inner sum vanishes for each n. For the second sum we obtain, using the partial fraction expansion for the cotangent:

$$\sum_{k=1}^{\infty}\sum_{n=1}^{\infty}(h_{n,k}+h_{n,-k})$$

$$= \sum_{k=1}^{\infty}\frac{1}{\tau}\cdot\sum_{n=1}^{\infty}\left(\frac{1}{n+\frac{k-1}{\tau}}+\frac{1}{-n+\frac{k-1}{\tau}}-\frac{1}{n+\frac{k}{\tau}}-\frac{1}{-n+\frac{k}{\tau}}\right)$$

$$= -\frac{\pi}{\tau}\cot\left(\frac{\pi}{\tau}\right)+1+\sum_{k=2}^{\infty}\left(\frac{\pi}{\tau}\cot\left(\pi\frac{k-1}{\tau}\right)-\frac{1}{k-1}-\frac{\pi}{\tau}\cot\left(\pi\frac{k}{\tau}\right)+\frac{1}{k}\right)$$

$$= \lim_{k\to\infty}\left(-\frac{\pi}{\tau}\cot\left(\pi\frac{k}{\tau}\right)+\frac{1}{k}\right)$$

$$= -\frac{\pi i}{\tau}\quad\text{(for } \tau\in\mathfrak{h}\text{).}$$

Definition: *The Dedekind η-function is defined as*

$$\eta(\tau) := q^{1/24}\cdot\prod_{n=1}^{\infty}(1-q^n).$$

Lemma 4.9: *We have*

$$\frac{d}{d\tau}\log\left(\eta(\tau)\right) = -2\pi i G_2(\tau), \tag{12}$$

$$\eta(-1/\tau) = e^{-2\pi i/8}\sqrt{\tau}\cdot\eta(\tau), \tag{13}$$

$$\eta(\tau+1) = e^{2\pi i/24}\eta(\tau). \tag{14}$$

In (13), it is understood that the root is to be taken so that the result again lies in $\mathfrak{h}$.

Proof:

$$\frac{d}{d\tau} \log(\eta(\tau)) = -2\pi i\left(-\frac{1}{24} + \sum_{n=1}^{\infty} \frac{nq^n}{1-q^n}\right)$$

$$= -2\pi i\left(-\frac{1}{24} + \sum_{n=1}^{\infty} \sigma_1(n)q^n\right),$$

which is precisely $-2\pi i G_2(\tau)$. From this and Lemma 4.8 we conclude that

$$d(\log(\eta(-1/\tau))) = -2\pi i G_2(-1/\tau)\frac{d\tau}{\tau^2}$$

$$= -2\pi i(G_2(\tau) - 1/(4\pi i \tau))\,d\tau$$

$$= d\log(\eta(\tau)) + 1/(2\tau)\,d\tau.$$

Therefore there is a constant k so that $\eta(-1/\tau) = k \cdot \tau^{1/2}\eta(\tau)$ for all $\tau \in \mathfrak{h}$. For $\tau = i$ this gives at once $k = e^{-\pi i/4}$. The equation (14) is obvious. $\qquad\square$

Corollary 4.10: *For the discriminant $\Delta(\tau)$ defined in Remark 3.7 we have*

$$\Delta(\tau) = \eta(\tau)^{24} = q \cdot \prod_{n=1}^{\infty} (1-q^n)^{24}.$$

Proof: η^{24} is, by (13) and (14), a modular form for the full modular group of weight 12, without zeroes in $\mathfrak{h}$ and with a zero of minimal order at ∞. Therefore $\Delta(\tau)/\eta(\tau)^{24} \in M_0(\mathrm{SL}_2(\mathbb{Z}))$ is constant by Corollary 4.2; since $\Delta(q) = q + O(q^2)$, this constant is 1. $\qquad\square$

Corollary 4.11: *The modular forms ε and $\widetilde{\varepsilon}$ of Remark 3.7 satisfy:*

$$\varepsilon = \frac{1}{16} \cdot \frac{\eta(\tau)^{16}}{\eta(2\tau)^8} \quad \text{and} \quad \widetilde{\varepsilon} = \frac{\eta(2\tau)^{16}}{\eta(\tau)^8}. \tag{15}$$

Proof: As already remarked in the proof of Corollary 4.10, we have $\mathrm{ord}_\infty(\Delta) = 1$ and therefore (since Δ is invariant under $\mathrm{SL}_2(\mathbb{Z})$) $\mathrm{ord}_0(\Delta) = 1$. For the functions $\xi(\tau) := \eta(\tau)^8$ and $\widetilde{\xi}(\tau) := \eta(2\tau)^8$ we therefore have:

$$\mathrm{ord}_\infty(\xi) = 1/3, \quad \mathrm{ord}_\infty(\widetilde{\xi}) = 2/3,$$

$$\mathrm{ord}_0(\xi) = 1/3, \quad \mathrm{ord}_0(\widetilde{\xi}) = 1/6.$$

Therefore both right sides in (15) are regular at the cusps. Because η has no zeroes in $\mathfrak{h}$, the right hand side of the asserted identities define holomorphic functions on $\mathfrak{h}$. To check their transformation law under $\Gamma_0(2)$ note that $\Gamma_0(2)$ is generated by $T = \left(\begin{smallmatrix} 1 & 1 \\ 0 & 1 \end{smallmatrix}\right)$ and $ST^{-2}S = -\left(\begin{smallmatrix} 1 & 0 \\ 2 & 1 \end{smallmatrix}\right)$ with $S = \left(\begin{smallmatrix} 0 & -1 \\ 1 & 0 \end{smallmatrix}\right)$. By Lemma 4.9 we have:

$$\xi|_4 T = e^{2\pi i/3}\xi,$$
$$\widetilde{\xi}|_4 T = e^{4\pi i/3}\widetilde{\xi},$$
$$\xi|_4 ST^{-2}S = e^{-4\pi i/3}\xi,$$
$$\widetilde{\xi}|_4 ST^{-2}S = \xi|_4 ST^{-1}S\left(\begin{smallmatrix} \sqrt{2} & 0 \\ 0 & 1/\sqrt{2} \end{smallmatrix}\right) = e^{-2\pi i/3}\xi|_4\left(\begin{smallmatrix} \sqrt{2} & 0 \\ 0 & 1/\sqrt{2} \end{smallmatrix}\right)$$
$$= e^{-2\pi i/3}\widetilde{\xi}.$$

Therefore both expressions above are modular forms for $\Gamma_0(2)$ of weight 4, and a comparison of the first two coefficients of the Fourier expansions with those of ε and $\widetilde{\varepsilon}$ yields the assertion. $\boxdot$

5 Theta functions, divisors, and elliptic functions

5.1 The Weierstraß σ-function

In the following sections we shall construct and investigate various explicit examples of elliptic functions (or rather theta functions or Jacobi forms). These will usually be given by infinite sums or products of simple analytic functions. We tacitly assume that the reader will verify that the products or sums converge in an appropriate way so that the functions defined by them have the required properties. As technical tool for this the following two facts are in general sufficient. Their proof can be found in any introductory book on complex analysis.

1) Let $\sum_{n\in\mathbb{N}} f_n(x)$ be a normally convergent series of meromorphic functions (i.e. for all compacta K there exists an $N \in \mathbb{N}$ such that the function f_n has no poles on K for $n > N$ and $\sum_{n>N} \sup_{x\in K} |f_n(x)| < \infty$). Then there exists a meromorphic function $f(x)$ with $\sum_{n\in\mathbb{N}} f_n(x) = f(x)$, $\{x \mid x \text{ pole of } f\} \subset \bigcup_n \{x \mid x \text{ pole of } f_n\}$, and at each pole x of f we have $\mathrm{ord}_x(f) \geq \min_{n\in\mathbb{N}}\{\mathrm{ord}_x(f_n)\}$.

2) Let the functions $u_n(x)$ be holomorphic on $\mathbb{C}$ and let $\prod_{n\in\mathbb{N}}(1 + u_n(x))$ be normally convergent (i.e. $\sum_{n\in\mathbb{N}} u_n(x)$ is normally convergent). Then a holomorphic limit value $u(x) = \prod_{n\in\mathbb{N}}(1 + u_n(x))$ exists. For this function u holds: $\{x \mid u(x) = 0\} = \bigcup_{n\in\mathbb{N}} \{x \mid 1 + u_n(x) = 0\}$, and at the zeroes x of u we have $\mathrm{ord}_x(u) = \sum_{n\in\mathbb{N}} \mathrm{ord}_x(1 + u_n)$.

In order to explicitly construct, for a prescribed divisor on an elliptic curve $\mathbb{C}/L$, an elliptic function with precisely this divisor, one needs the Weierstraß σ-function. In turn, one obtains the best understanding of this by considering theta functions.

Definition: *Let $L \subset \mathbb{C}$ be a lattice. A meromorphic function f is called a **theta function** for L if $\frac{d^2}{dx^2} \log(f(x))$ is elliptic with respect to L. The theta functions defined by $f(x) = \exp\left(ax^2 + bx + c\right)$ for $a, b, c \in \mathbb{C}$ are called **trivial theta functions**.*

Remarks:

1) The theta functions different from zero form a group with respect to multiplication.

2) A meromorphic function f is a theta function with respect to the lattice L if and only if for each $\gamma \in L$, $f(x + \gamma)/f(x)$ is a trivial theta function of the form $x \mapsto \exp(bx + c)$. In particular, for a theta function f its divisor

$$\mathrm{div}\,(f) := \sum_{x+L \in \mathbb{C}/L} \mathrm{ord}_{x+L}\,(f) \cdot (x + L)$$

is therefore well-defined. By abuse of language we sometimes write (x) for $(x + L)$ when dealing with divisors on the quotient $\mathbb{C}/L$.

With the notations $\mathrm{Div}\,(E) = \{$divisors on $E = \mathbb{C}/L\}$, $\Theta_L = \{$theta functions $\neq 0$ for $L\}$ and $\widetilde{\Theta}_L = \{$trivial theta functions for $L\}$ we have the following

Theorem 5.1: *For a fixed lattice L, the canonical assignment $f \mapsto \mathrm{div}\,(f)$ induces an exact sequence of groups:*

$$1 \to \widetilde{\Theta}_L \hookrightarrow \Theta_L \to \mathrm{Div}\,(E) \to 1.$$

Proof: Clearly $f \mapsto \mathrm{div}\,(f)$ is a homomorphism. Let f lie in the kernel of $\Theta_L \to \mathrm{Div}\,(E)$, i.e. f has neither zeroes nor poles, hence $g := \log(f)$ is globally defined, so $f = \exp(g)$. Since $\frac{d^2 g}{dx^2}$ is elliptic, it follows that $\frac{d^2 g}{dx^2} = 2a$, i.e. $g = ax^2 + bx + c$ for suitable $a, b, c \in \mathbb{C}$, hence f is a trivial theta function. We shall see in Theorem 5.2 that the divisor (0) has a preimage, namely the function σ. The surjectivity of $\Theta_L \to \mathrm{Div}\,(E)$ now follows at once, since for $D = \sum_{i=1}^{N} n_i \cdot (P_i) \in \mathrm{Div}\,(E)$ with $x_i \equiv P_i\,(L)$ we have

$$\mathrm{div}\left(\prod_{i=1}^{N} \sigma(x - x_i)^{n_i}\right) = D. \qquad \square$$

The function σ used above does indeed exist. Without loss of generality, we consider lattices of the form $2\pi i(\mathbb{Z}\tau + \mathbb{Z})$, for $\tau \in \mathfrak{h}$:

Theorem 5.2: *For each $\tau \in \mathfrak{h}$, the product*

$$\sigma(\tau, x) := x \prod_{\substack{\gamma \in 2\pi i(\mathbb{Z}\tau + \mathbb{Z}) \\ \gamma \neq 0}} \left(1 - \frac{x}{\gamma}\right) \cdot \exp\left(\frac{x}{\gamma} + \frac{1}{2}\left(\frac{x}{\gamma}\right)^2\right) \qquad (1)$$

*is normally convergent in $x \in \mathbb{C}$ and defines a theta function for the lattice $2\pi i(\mathbb{Z}\tau + \mathbb{Z})$
with $\mathrm{div}\,(\sigma(\tau, \cdot)) = (0)$. In terms of the Eisenstein series $G_{2k}(\tau)$ of §3, equation (3),
one has*

$$\sigma(\tau, x) = \exp\left(G_2(\tau) \cdot x^2\right) \cdot \left(\zeta^{1/2} - \zeta^{-1/2}\right) \cdot \prod_{n=1}^{\infty} \frac{(1 - q^n\zeta)(1 - q^n\zeta^{-1})}{(1 - q^n)^2} \qquad (2)$$

$$= x \cdot \exp\left(-\sum_{k=2}^{\infty} \frac{2}{(2k)!} G_{2k}(\tau) \cdot x^{2k}\right). \qquad (3)$$

Proof: We have

$$\frac{d}{dx} \log(\sigma) = \frac{1}{x} + \sum_{\substack{\gamma \in 2\pi i(\mathbb{Z}\tau + \mathbb{Z}) \\ \gamma \neq 0}} \left(\frac{1}{x - \gamma} + \frac{1}{\gamma} + \frac{x}{\gamma^2}\right)$$

and therefore $\frac{d^2}{dx^2} \log(\sigma) = -\wp(\tau, x)$. Hence σ is a theta function. The assertion
about the zeroes is also clear, since by what was proved above the product is normally
convergent.

On (2): For $|q| < \min\{|\zeta|, |\zeta^{-1}|\}$ we have

$$\frac{d^2}{dx^2} \log(\text{right side of (2)})$$

$$= 2G_2(\tau) + \frac{d}{dx}\left(\frac{1}{2}\coth\left(\frac{x}{2}\right) + \sum_{n=1}^{\infty}\left(\frac{-q^n\zeta}{1 - q^n\zeta} + \frac{q^n\zeta^{-1}}{1 - q^n\zeta^{-1}}\right)\right)$$

$$= 2G_2(\tau) - \frac{1}{4\sinh\left(\frac{x}{2}\right)^2} + \frac{d}{dx}\left(\sum_{n=1}^{\infty}\sum_{k=1}^{\infty}(\zeta^{-k} - \zeta^k) \cdot q^{nk}\right)$$

$$= 2G_2(\tau) - \frac{1}{\left(\zeta^{1/2} - \zeta^{-1/2}\right)^2} - \sum_{n=1}^{\infty}\sum_{k=1}^{\infty} k(\zeta^k + \zeta^{-k}) \cdot q^{nk}$$

$$= \frac{-1}{12} + 2\sum_{n=1}^{\infty} \sigma_1(n)q^n - \frac{1}{\left(\zeta^{1/2} - \zeta^{-1/2}\right)^2} - \sum_{n=1}^{\infty}\sum_{d|n} d(\zeta^d + \zeta^{-d})q^n$$

$$= -\wp(\tau, x),$$

the latter by Lemma 3.3 and the definition of G_2. Since also $\frac{d^2}{dx^2} \log(\sigma) = -\wp(\tau, x)$, as proved above, we have

$$\sigma(\tau, x) = \exp\left(b(\tau)x + c(\tau)\right) \cdot (\text{right side of (2)}).$$

In addition, σ begins with $x + O(x^2)$ and is odd. The right side of (2)
likewise begins with $x + O(x^2)$ and is odd. Hence, $b = c = 0$.

On (3):

$$\log\left(\text{right side of (1)}\right) - \log\left(x\right) = \sum_{\gamma \neq 0}\left(-\sum_{n=1}^{\infty}\frac{x^n}{n\gamma^n} + \frac{x}{\gamma} + \frac{1}{2}\left(\frac{x}{\gamma}\right)^2\right)$$

$$= -\sum_{\gamma \neq 0}\sum_{n=3}^{\infty}\frac{x^n}{n\gamma^n}$$

$$= -\sum_{k=2}^{\infty}\frac{1}{2k}\,x^{2k}\sum_{\substack{\gamma \in 2\pi i(\mathbb{Z}\tau + \mathbb{Z}) \\ \gamma \neq 0}}\frac{1}{\gamma^{2k}}$$

$$= -\sum_{k=2}^{\infty}\frac{1}{2k}\,x^{2k}\,\frac{2}{(2k-1)!}\,G_{2k}(\tau)$$

$$= \log\left(\text{formula (3)}\right) - \log\left(x\right),$$

where for the next to last equality we have used the identity (4) from §3. Therefore, (2) and (3) are equal up to multiplication by a constant, which is obviously equal to 1. $\quad\square$

Sometimes it is better to consider the following theta function, with the same divisor (0), in place of the σ-function; this theta function is not normalized by $x + O(x^5)$, but instead has a more conveniently representable behavior under translation by lattice vectors.

Corollary 5.3: *For each* $\tau \in \mathfrak{h}$,

$$\Phi(\tau, x) := \sigma(\tau, x) \cdot \exp\left(-G_2(\tau) \cdot x^2\right)$$

is a theta function for the lattice $2\pi i(\mathbb{Z}\tau + \mathbb{Z})$ *with* $\mathrm{div}\left(\Phi(\tau, \cdot)\right) = (0)$ *and the normalization* $\Phi(\tau, x) = x + O(x^3)$. *We have thereby:*

$$\Phi(\tau, x) = (\zeta^{1/2} - \zeta^{-1/2}) \cdot \prod_{n=1}^{\infty}\frac{(1 - q^n\zeta)(1 - q^n\zeta^{-1})}{(1 - q^n)^2} \tag{4}$$

$$= x \cdot \exp\left(-\sum_{k=1}^{\infty}\frac{2}{(2k)!}\,G_{2k}(\tau) \cdot x^{2k}\right) \tag{5}$$

and

$$\frac{d^2}{dx^2}\log\left(\Phi\right) = -\wp(\tau, x) - 2G_2(\tau) = -\sum_{n \in \mathbb{Z}}\frac{1}{4\sinh\left(\frac{\pi i n\tau + x}{2}\right)^2}. \qquad \square$$

The σ- and Φ-functions exhibit interesting invariance properties:

Theorem 5.4: *We have*

$$\sigma\left(A\tau, \tfrac{x}{c\tau + d}\right) \cdot (c\tau + d) = \sigma(\tau, x), \tag{6}$$

$$\Phi\left(A\tau, \tfrac{x}{c\tau + d}\right) \cdot (c\tau + d) = \exp\left(\tfrac{cx^2}{4\pi i(c\tau + d)}\right) \cdot \Phi(\tau, x), \tag{7}$$

for all $A = \begin{pmatrix} a & b \\ c & d \end{pmatrix} \in SL_2(\mathbb{Z})$, *and*

$$\Phi(\tau, x + 2\pi i \cdot (\lambda\tau + \mu)) = q^{-\frac{\lambda^2}{2}} e^{-\lambda x}(-1)^{\lambda+\mu} \cdot \Phi(\tau, x), \tag{8}$$

for all $\lambda, \mu \in \mathbb{Z}$.

Proof: Equation (6) follows directly from the definition of the σ-function. (7) follows from (6) and $G_2(A\tau)(c\tau + d)^{-2} = G_2(\tau) - \frac{c}{4\pi i(c\tau+d)}$ (by Lemma 4.8). One verifies (8) using (4), first for $(\lambda, \mu) = (1, 0)$, resp. $(0, \mu)$, and then by induction on λ (using $1 + 3 + \cdots + (2\lambda - 1) = \lambda^2$). $\qquad\qquad$ ⌘

5.2 The Jacobi form φ

Theorem 5.5: *Let* $L \subset \mathbb{C}$ *be a lattice,* K_L *the field of elliptic functions with respect to* L, $\mathrm{Div}_0(E)$ *the group of divisors of degree* 0 *on the curve* $E = \mathbb{C}/L$ *and* $\mathrm{Div}_p(E)$ *the subgroup of principal divisors on* E. *Then* $f \mapsto \mathrm{div}(f)$ *induces an exact sequence*

$$1 \to \mathbb{C}^* \to K_L^* \to \mathrm{Div}_p(E) \to 0 \tag{9}$$

and we have

$$\mathrm{Div}_p(E) = \{D \in \mathrm{Div}_0(E) \mid D = \sum_{P\in\mathbb{C}/L} n_P \cdot (P), \quad \sum_{P\in\mathbb{C}/L} n_P \cdot P \equiv 0\ (L)\}.$$

Proof: The exactness of the sequence (9) is immediate except for the exactness at the middle place. The latter is equivalent to the fact that there is no non-constant elliptic function without poles (recall that any doubly periodic holomorphic function is bounded on a fundamental mesh, hence on $\mathbb{C}$, hence constant by Liouville's theorem).
The inclusion " $\subset$ " of the asserted identity follows from the easily proved identities

$$\sum_{P\in\mathbb{C}/L} \mathrm{ord}_P(f) = \oint d\log(f) = 0,$$

$$\sum_{P\in\mathbb{C}/L} \mathrm{ord}_P(f) \cdot P = \oint x\, d\log(f) \equiv 0\ (L),$$

where the integrals are taken along the boundary of a suitably chosen (not passing through zeroes or poles) fundamental mesh of L. For the inclusion " $\supset$ ": If $D = \sum_{i=1}^{N} n_i \cdot (x_i) \in \mathrm{Div}_0(E)$ (with not necessarily distinct x_i and $n_i \in \{\pm 1\}$) such that $\sum_{i=1}^{N} n_i x_i \in L$, then one can assume $\sum_{i=1}^{N} n_i x_i = 0$ (otherwise, replace x_1 by $x_1 - n_1 \sum_{i=1}^{N} n_i x_i$). Then the product $f(x) := \prod_{i=1}^{N} \sigma(x - x_i)^{n_i}$ is elliptic with divisor D, since for $\gamma \in L$ we have

$$\prod_{i=1}^{N} \frac{\sigma(x - x_i + \gamma)^{n_i}}{\sigma(x - x_i)^{n_i}} = \prod_{i=1}^{N} \exp\left(b(x - x_i)n_i + cn_i\right)$$

$$= \exp\left((bx + c)\sum_{i=1}^{N} n_i - b\sum_{i=1}^{N} n_i x_i\right) = 1$$

where b and c are determined by $\sigma(x + \gamma)/\sigma(x) = \exp(bx + c)$.   ⌸

We shall use Theorem 5.5 in the following to give an axiomatic characterization of various elliptic functions.

Theorem 5.6: *For $\tau \in \mathfrak{h}$, let $\varphi(\tau, x)$ denote the elliptic function with respect to x for the lattice $L = 2\pi i(\mathbb{Z} \cdot 2\tau + \mathbb{Z})$ with the divisor $(\pi i) + (\pi i \cdot (1 + 2\tau)) - (\pi i \cdot 2\tau) - (0)$ and the normalization $\varphi(\tau, x) = \frac{1}{x} + O(1)$ for $x \to 0$ (which, according to Theorem 5.5 exists and is uniquely determined). Then we have:*

$$\varphi\left(\tfrac{a\tau+b}{c\tau+d}, \tfrac{x}{c\tau+d}\right) \cdot (c\tau + d)^{-1} = \varphi(\tau, x) \tag{10}$$

for all $A = \begin{pmatrix} a & b \\ c & d \end{pmatrix} \in \Gamma_0(2)$,

$$\varphi(\tau, x)^2 = \wp(\tau, x) - e_1(\tau), \tag{11}$$

$$\varphi(\tau, x) = \frac{1}{2} \frac{\zeta^{1/2} + \zeta^{-1/2}}{\zeta^{1/2} - \zeta^{-1/2}} \cdot \prod_{n=1}^{\infty} \frac{(1 + q^n\zeta)(1 + q^n\zeta^{-1})(1 - q^n)^2}{(1 - q^n\zeta)(1 - q^n\zeta^{-1})(1 + q^n)^2}, \tag{12}$$

$$\varphi(\tau, x) = \lim_{N \to \infty} \left(\sum_{n=-N}^{N} \frac{(-1)^n}{2} \coth\left(\frac{2\pi i n\tau + x}{2}\right)\right), \tag{13}$$

$$\varphi(\tau, x) = \frac{\Phi(\tau, x - \pi i)}{\Phi(\tau, x)\Phi(\tau, -\pi i)} = \frac{\sigma(\tau, x - \pi i)}{\sigma(\tau, x)\sigma(\tau, -\pi i)} \exp(2\pi i x \cdot G_2(\tau)). \tag{14}$$

Proof:

On (10):   Choose a fixed τ, and define

$$\psi(\tau, x) := \varphi\left(\tfrac{a\tau+b}{c\tau+d}, \tfrac{x}{c\tau+d}\right) \cdot (c\tau + d)^{-1} = \frac{1}{x} + O(1).$$

For $\lambda, \mu \in \mathbb{Z}$ we have

$$\begin{aligned}
\psi(\tau, x + 2\pi i(2\lambda\tau + \mu)) &= \varphi\left(A\tau, \tfrac{x+2\pi i(2\lambda\tau+\mu)}{c\tau+d}\right) \cdot (c\tau + d)^{-1} \\
&= \varphi\left(A\tau, \tfrac{x}{c\tau+d} + 2\pi i(\lambda' \cdot 2A\tau + \mu')\right) \cdot (c\tau + d)^{-1} \\
&= \varphi\left(A\tau, \tfrac{x}{c\tau+d}\right) \cdot (c\tau + d)^{-1} \\
&= \psi(\tau, x),
\end{aligned}$$

where $(\lambda' \; \mu')\begin{pmatrix} a & 2b \\ \frac{c}{2} & d \end{pmatrix} = (\lambda \; \mu)$. Furthermore, ψ has the same zeroes and poles as φ, as one easily sees. Since φ is uniquely determined by these properties, we therefore have

$$\varphi(\tau, x) = \psi(\tau, x) = \varphi\left(A\tau, \tfrac{x}{c\tau+d}\right)(c\tau + d)^{-1}.$$

On (11): $\varphi(\tau, x + 2\pi i\tau) = -\varphi(\tau, x)$, since $\varphi(\tau, x + 2\pi i\tau) + \varphi(\tau, x)$ is elliptic with respect to $\widetilde{L} = 2\pi i(\mathbb{Z}\tau + \mathbb{Z})$ and has at most one pole (of order one) on $\widetilde{L}$, and is therefore constant. The value of the constant is 0, as one sees by taking $x = \pi i$. Hence $\varphi(\tau, x)^2$ is elliptic with respect to $\widetilde{L}$ and has the same poles of order two as $\wp(\tau, x)$. Thus we have $\varphi^2(\tau, x) = \wp(\tau, x) + \text{const.}$ Setting $x = \pi i$ gives $\text{const.} = -\wp(\tau, \pi i) = -e_1(\tau)$.

On (12): Call the right side $p(x)$. The function p is meromorphic and the various factors yield for the divisor: The denominator of $\coth(x)$ yields poles at $2\pi i(0\tau + \mathbb{Z})$, and the numerator yields zeroes at $2\pi i(0\tau + \mathbb{Z}) + \pi i$. The product gives zeroes at $2\pi i(N\tau + \mathbb{Z}) + \pi i$, resp. $2\pi i(-N\tau + \mathbb{Z}) + \pi i$, and poles at $2\pi i(N\tau + \mathbb{Z})$, resp. $(-N\tau + \mathbb{Z})$. The divisors of φ and p therefore coincide. The invariance under $x \mapsto x + 2\pi i$ is clear; one checks the behavior under $x \mapsto x + 2\pi i\tau$ (hence $\zeta \mapsto q\zeta$) as follows:

$$p(x + 2\pi i\tau) = \frac{1}{2}\frac{q\zeta + 1}{q\zeta - 1} \cdot \prod_{n=1}^{\infty} \frac{\left(1 + q^{n+1}\zeta\right)\left(1 + q^{n-1}\zeta^{-1}\right)(1 - q^n)^2}{(1 - q^{n+1}\zeta)(1 - q^{n-1}\zeta^{-1})(1 + q^n)^2}$$
$$= -p(x).$$

The normalization of $p(x)$ is likewise $\frac{1}{x} + O(1)$. Hence $p(x)$ has precisely those properties which characterize φ uniquely, i.e. $p = \varphi$.

On (13): Let $s(x)$ denote the function defined by the sum. Obviously $s(x + 2\pi i) = s(x)$, and one can easily show that also $s(x + 2\pi i\tau) = -s(x)$. The poles of s have order 1 and lie at $2\pi i(\mathbb{Z}\tau + \mathbb{Z})$. Moreover, $s(x) = \frac{1}{x} + O(1)$. Hence $\varphi = s + \text{const.}$ But s is odd, as is φ obviously $(-\varphi(\tau, -x)$ satisfies the same conditions as $\varphi(\tau, x))$, hence the constant must be 0.

On (14): This follows again by verification of the properties characterizing φ for the right side of (14), by means of the various properties of Φ given in the previous section.

Note that (12) can also be deduced from (14) and (4). Similarly, one could derive (10) from (14) and the transformation law of Φ derived in the previous section.

Exercise (expansion of φ at the cusp 0):

$$\varphi^0(\tau, x) := \varphi\left(\tfrac{-1}{\tau}, \tfrac{x}{\tau}\right) \cdot \tau^{-1}$$

is elliptic with respect to the lattice $\widetilde{\widetilde{L}} = 2\pi i(\mathbb{Z}\tau + \mathbb{Z} \cdot 2)$ and has the expansion $\frac{1}{x} + O(1)$ for $x \to 0$. The poles are at $\widetilde{\widetilde{L}}$ and $2\pi i + \widetilde{\widetilde{L}}$, and we have

$$\varphi^0(\tau, x) = \frac{1}{\zeta^{1/2} - \zeta^{-1/2}} \cdot \prod_{n=1}^{\infty} \left(\frac{\left(1 - q^{n/2}\right)^2}{\left(1 - q^{n/2}\zeta\right)\left(1 - q^{n/2}\zeta^{-1}\right)}\right)^{(-1)^n} \tag{15}$$

$$= \frac{1}{2} \cdot \sum_{n \in \mathbb{Z}} \frac{1}{\sinh\left(\frac{x + 2\pi i n\tau}{2}\right)} \tag{16}$$

$$= \sum_{n \in \mathbb{Z}} \frac{1}{q^{n/2}\zeta^{1/2} - q^{-n/2}\zeta^{-1/2}} \tag{17}$$

$$= \frac{\Phi(\tau, x - \pi i \tau)}{\Phi(\tau, x)\Phi(\tau, -\pi i \tau)} \, e^{-x/2}, \tag{18}$$

(for (16) verify that the right side s has the same poles and residues as φ^0, whence $\varphi^0 = s + \mathrm{const.}$, and use that φ^0 and s are odd functions).

6   Elliptic genera of level N

6.1   Tools: The Φ-function

One obtains explicit formulas in the theory of elliptic genera of level N by application of the definitions and theorems of §5, in the same way this was done there in Theorem 5.6 for $N = 2$. From consideration of formula (3) in §5

$$\sigma(\tau, x) = x \cdot \exp\!\left(-\sum_{k=2}^{\infty} \frac{2}{(2k)!} \, G_{2k}(\tau) \cdot x^{2k}\right)$$

we are led at once to the question: How do the modular forms G_k enter into these formulas? The answer is obtained from Theorem 3.1, which we recall here in a reformulation suitable to our aims.

Theorem 6.1 *For each $\tau \in \mathfrak{h}$, let $\Psi(\tau, x)$ be a doubly periodic function in x with period lattice $L = 2\pi i(\mathbb{Z}\tau + \mathbb{Z})$. Suppose that for a suitable number $k \in \mathbb{Z}$, $\Gamma \subset \mathrm{SL}_2(\mathbb{Z})$, and group homomorphism $\varepsilon : \Gamma \to \mathbb{C}^*$ we have*

$$\Psi\!\left(A\tau, \tfrac{x}{c\tau+d}\right)(c\tau + d)^{-k} = \varepsilon(A) \cdot \Psi(\tau, x)$$

for all $A = \left(\begin{smallmatrix} a & b \\ c & d \end{smallmatrix}\right) \in \Gamma$ (briefly: Ψ is invariant under Γ of weight k with Nebentyp ε). Let $\omega = 2\pi i(\alpha\tau + \beta)$ be an N-division point of L and $a_n(\tau)$ the n-th coefficient of the Laurent expansion at the point ω. Then we have:

$$a_n(A\tau)(c\tau + d)^{-(n+k)} = \varepsilon(A)a_n(\tau)$$

for all $A = \left(\begin{smallmatrix} a & b \\ c & d \end{smallmatrix}\right) \in \Gamma'$ with $\Gamma' = \{A \mid (\alpha, \beta)A \equiv (\alpha, \beta) \, (N)\}$; in particular, for $\omega = 2\pi i/N$ we have $\Gamma' = \Gamma_1(N) \cap \Gamma$.   ▱

If $A \in \mathrm{SL}_2(\mathbb{Z})$, then $\sigma\!\left(A\tau, \tfrac{x}{c\tau+d}\right) \cdot (c\tau + d)$ is again a theta function for the lattice $2\pi i(\mathbb{Z}\tau + \mathbb{Z})$ with divisor $(0 + 2\pi i(\mathbb{Z}\tau + \mathbb{Z}))$, hence equal to a trivial theta function times $\sigma(\tau, x)$. Indeed, as we have seen in formula (6) from §5, $\sigma(\tau, x)$ is invariant under $SL_2(\mathbb{Z})$. Therefore σ'/σ is also invariant under $SL_2(\mathbb{Z})$ (of weight 1), hence the k-th Taylor coefficient of $\frac{d}{dx}\log(\sigma)$ is invariant under $SL_2(\mathbb{Z})$ (of weight $k+1$), and in fact it is a multiple of G_{k+1} for odd k and 0 for even k.

In view of Theorem 6.1, a formula analogous to formula (5) from §5, giving an expansion of Φ at a general N-division point, is of interest:

Theorem 6.2: *Let $L = 2\pi i(\mathbb{Z}\tau + \mathbb{Z})$ and let $\omega = 2\pi i(\alpha\tau + \beta)$ with $-1 < \alpha, \beta < 1$. Then*

$$\Phi(\tau, x) = \exp\left(-2\sum_{k=1}^{\infty} \frac{(x-\omega)^k}{k!} G_k^{(\omega)}(\tau)\right) \cdot \begin{cases} x, & \text{if } \omega = 0, \\ \Phi(\tau, \omega), & \text{otherwise.} \end{cases} \tag{1}$$

Here

$$G_k^{(\omega)}(\tau) = -\frac{c_k^{(\omega)}}{2k} + \sum_{n=1}^{\infty}\left(\sum_{d\mid n} \frac{e^{d\omega} + (-1)^k e^{-d\omega}}{2} d^{k-1}\right) \cdot q^n, \tag{2}$$

where the $c_k^{(\omega)}$ are defined by:

$$\frac{1}{2}\coth\left(\frac{x}{2}\right) = \sum_{k=0}^{\infty} \frac{c_k^{(\omega)}}{k!}(x-\omega)^{k-1}.$$

Proof: From (4) of Corollary 5.3 we obtain for $|q| < |\zeta|, |\zeta|^{-1}$, i.e. for $|\mathrm{Im}\,(x)| < 2\pi\,\mathrm{Im}\,(\tau)$,

$$\frac{\partial}{\partial x}\log(\Phi(\tau, x)) = \frac{1}{2}\coth\left(\frac{x}{2}\right) - \sum_{n=1}^{\infty}\sum_{d\mid n}(e^{dx} - e^{-dx})q^n.$$

Writing

$$e^{dx} - e^{-dx} = \sum_{k=0}^{\infty}(e^{d\omega} + (-1)^{k+1}e^{-d\omega})d^k \frac{(x-\omega)^k}{k!},$$

we find

$$\frac{\partial}{\partial x}\log(\Phi(\tau, x)) = \frac{1}{2}\coth\left(\frac{x}{2}\right) - 2\sum_{k=0}^{\infty} \widetilde{G_{k+1}^{(\omega)}}(\tau)\frac{(x-\omega)^k}{k!}$$

for x near ω (note that by assumption $|\mathrm{Im}\,(x)| < 2\pi\,\mathrm{Im}\,(\tau)$, i.e. ω is in the domain valid for the cited expansion of $\frac{\partial}{\partial x}\log(\Phi(\tau, x))$). Here $\widetilde{G_{k+1}^{(\omega)}} = G_{k+1}^{(\omega)} - \text{const.}$ From the latter equation the claimed formula can now easily be deduced. $\qquad\square$

Remarks:

1) If $\omega \in \frac{1}{N}L$ then it can easily be verified that the constant term of $G_k^{(\omega)}(\tau)$ is given by

$$-\frac{c_k^{(\omega)}}{2k} = \sum_{\nu=0}^{N-1} \frac{B_k\left(\frac{\nu}{N}\right)e^{\omega\nu}}{2k} - \delta_{k,1} \cdot \frac{1}{4}$$

($\delta_{k,1} = 0$ or 1 accordingly as $k \neq 1$ or $k = 1$). Here the $B_k(x)$ denote the Bernoulli polynomials, i.e.

$$\frac{te^{xt}}{e^t - 1} = \sum_{k=0}^{\infty} \frac{B_k(x)}{k!} t^k .$$

2)   If $\omega \in \frac{1}{N}L$, $\omega = 2\pi i \frac{\alpha\tau+\beta}{N}$, then, by Theorem 6.1 applied to $\frac{\partial}{\partial x} \log(\Phi(\tau,x))$, the function $G_k^{(\omega)}(\tau)$ transforms like an element of $M_k(\Gamma)$, where $\Gamma = A^{-1}\Gamma_1(N)A$ with A denoting any matrix with lower row equal to (α,β). In fact, it is easily checked that $G_k^{(\omega)} \in M_k(\Gamma)$: The regularity at the cusps follows from the fact that $G_k^{(\omega)}|_k M = G_k^{(\omega')}$ with a suitable ω' for any $M \in \mathrm{SL}_2(\mathbb{Z})$, as is easily deduced from the transformation laws of $\Phi(\tau,x)$ as stated in (7), (8) of Theorem 5.4.

Example: Applying the preceding remark 2) to $\omega = -2\pi i/3$ one finds

$$G_1^{(-2\pi i/3)}(\tau) = -\frac{\sqrt{-3}}{2} \cdot \left(\frac{1}{6} + \sum_{n=1}^{\infty} \sum_{d|n} \left(\tfrac{d}{3}\right) \cdot q^n \right) \in M_1(\Gamma_1(3)),$$

$$G_3^{(-2\pi i/3)}(\tau) = -\frac{\sqrt{-3}}{2} \cdot \left(-\frac{1}{486} + \sum_{n=1}^{\infty} \sum_{d|n} d^2 \left(\tfrac{d}{3}\right) \cdot q^n \right) \in M_3(\Gamma_1(3)).$$

Here $\left(\tfrac{d}{3}\right)$ is the Legendre symbol (i.e. $\left(\tfrac{d}{3}\right) = -1, 0, 1$ for $d \equiv -1, 0, 1$ (3)).

Remark: One can show that

$$M_*(\Gamma_1(3)) = \mathbb{C}[G_1^{(-2\pi i/3)}, G_3^{(-2\pi i/3)}].$$

6.2   The function f for elliptic genera of level N

We now want to use the methods presented above to study the function $f(x)$ introduced by Hirzebruch in Chapter 7.

Theorem 6.3: *Let $N \in \mathbb{N}$ and $\alpha, \beta \in \mathbb{Z}$, not both divisible by N. Then there is a unique function $f_{\alpha,\beta}(\tau, x)$ with the following properties:*

1)   *$x \mapsto f_{\alpha,\beta}(\tau, x)$ is a theta function on the lattice $2\pi i(\mathbb{Z}\tau + \mathbb{Z})$.*
2)   *The divisor of $f_{\alpha,\beta}(\tau, x)$ is $(0) - \left(2\pi i \frac{(\alpha\tau+\beta)}{N}\right)$.*
3)   *$f_{\alpha,\beta}(\tau, x) = x + O\left(x^2\right)$ for $x \to 0$ and for all $\lambda, \mu \in \mathbb{Z}$ there exists an N-th root of unity ν so that*

$$f_{\alpha,\beta}(\tau, x + 2\pi i(\lambda\tau + \mu)) = \nu \cdot f_{\alpha,\beta}(\tau, x),$$

i.e. $x \mapsto f_{\alpha,\beta}^N(\tau, x)$ is doubly periodic with respect to $2\pi i(\mathbb{Z}\tau + \mathbb{Z})$.

Remark: Obviously one has $f_{\alpha,\beta}(\tau,x) = f_{\alpha',\beta'}(\tau,x)$ if $(\alpha,\beta) \equiv (\alpha',\beta')\,(N)$.

Proof: By Theorem 5.1, $f_{\alpha,\beta}$ satisfying 1) and 2) certainly exists and is unique up to multiplication with e^{ax^2+bx+c}. Now $f^N_{\alpha,\beta}$ is a theta function with divisor $N \cdot \big((0) - (2\pi i \frac{\alpha\tau+\beta}{N})\big)$, and $N \cdot \big(0 - 2\pi i \frac{\alpha\tau+\beta}{N}\big) \in 2\pi i(\mathbb{Z}\tau + \mathbb{Z})$. By combining Theorem 5.1 and Theorem 5.5, we find that $f^N_{\alpha,\beta}$ is, up to multiplication by a trivial theta function θ, a doubly periodic function on $2\pi i(\mathbb{Z}\tau + \mathbb{Z})$. Multiplying $f_{\alpha,\beta}$ by const. $\cdot \sqrt[N]{\theta}$, if necessary, one can therefore assume that $f^N_{\alpha,\beta}$ is elliptic with respect to $2\pi i(\mathbb{Z}\tau + \mathbb{Z})$ and satisfies $f_{\alpha,\beta}(\tau,x) = x + O(x^2)$ $(x \to 0)$. Note that this $f^N_{\alpha,\beta}$ is uniquely determined (as is also implied by Theorem 5.5). This proves the theorem. $\qquad\boxed{}$

Theorem 6.4: *Let* $f_{\alpha,\beta}(\tau,x)$ *be as in the preceding theorem and let* $f(\tau,x) = f_{0,1}(\tau,x)$. *Then one has:*

(i)

$$f_{\alpha,\beta}(\tau,x) = \frac{\Phi(\tau,x)\Phi(\tau,-\omega)}{\Phi(\tau,x-\omega)} \cdot e^{\alpha x/N} \qquad (\omega = \tfrac{2\pi i(\alpha\tau+\beta)}{N}) \tag{3}$$

$$= 2e^{\frac{\alpha x}{N}}\,\frac{\sinh\left(\frac{x}{2}\right)\sinh\left(\frac{-\omega}{2}\right)}{\sinh\left(\frac{x-\omega}{2}\right)} \tag{4}$$

$$\prod_{n=1}^{\infty} \frac{(1-q^n e^x)(1-q^n e^{-x})(1-q^n e^{w})(1-q^n e^{-w})}{(1-q^n e^{x-\omega})(1-q^n e^{-x+\omega})(1-q^n)^2}, \tag{5}$$

$$f(\tau,x) = x \cdot \exp\left(2\sum_{k=1}^{\infty} f_k(\tau)\cdot\frac{x^k}{k!} - 2\sum_{\substack{k=1\\k\equiv 0\,(2)}}^{\infty} G_k(\tau)\cdot\frac{x^k}{k!}\right), \tag{6}$$

where

$$f_k(\tau) = -\frac{c_k^{(-2\pi i/N)}}{2k} + \sum_{n=1}^{\infty}\left(\sum_{d|n} d^{k-1}\cdot\begin{cases}\cos\left(\frac{-2\pi id}{N}\right) & \text{for } k \equiv 0\,(2)\\[4pt]\sin\left(\frac{-2\pi id}{N}\right) & \text{for } k \equiv 1\,(2)\end{cases}\right)\cdot q^n$$

$$\in M_k(\Gamma_1(N)), \tag{7}$$

with $c_k^{(\omega)}$ *as in Theorem 6.2.*

(ii)

$$f_{\alpha,\beta}\left(A\tau,\tfrac{x}{c\tau+d}\right)(c\tau+d) = f_{(\alpha,\beta)A}(\tau,x) \tag{8}$$

for all $A \in \mathrm{SL}_2(\mathbb{Z})$, *in particular for* $f(\tau,x)$ *one has*

$$f\left(A\tau,\tfrac{x}{c\tau+d}\right)(c\tau+d) = f(\tau,x) \quad \text{for all } A \in \Gamma_1(N). \tag{9}$$

(iii)

$$f_{\alpha,\beta}(\tau,x+2\pi i(\lambda\tau+\mu)) = \exp\left(2\pi i(\alpha\mu-\beta\lambda)/N\right)\cdot f_{(\alpha,\beta)}(\tau,x), \tag{10}$$

for all $\lambda,\,\mu \in \mathbb{Z}$.

Proof: The right side of (3) satisfies the conditions 1), 2), and the normalization condition in 3) of Theorem 6.3: One verifies (10) for the right side, using formula (8) from §5. Hence the right side of (3) also satisfies 3) in Theorem 6.3, and so is equal to $f_{\alpha,\beta}$. This also implies (10). The product representation of Φ then yields (5). Further, (6) is a consequence of (3) and Theorem 6.2 (formulas (1) and (2)). Formula (7) follows from (9) by logarithmic differentiation of f and Theorem 6.1. Formula (9) follows from (8), and (8) follows from the fact that $f_{\alpha,\beta}\big(A\tau, \frac{x}{c\tau+d}\big) \cdot (c\tau + d)$ satisfies the conditions 1), 2) and 3) of Theorem 6.3, with $(\alpha, \beta)A$ in place of (α, β): e.g., 3) results, with $(\lambda',\mu')A = (\lambda,\mu)$, as follows:

$$f_{\alpha,\beta}\Big(A\tau, \tfrac{x+2\pi i(\lambda\tau+\mu)}{c\tau+d}\Big) = f_{\alpha,\beta}\Big(A\tau, \tfrac{x+2\pi i\big(\lambda'(a\tau+b)+\mu'(c\tau+d)\big)}{c\tau+d}\Big)$$
$$= f_{\alpha,\beta}\Big(A\tau, \tfrac{x}{c\tau+d} + 2\pi i(\lambda'A\tau + \mu')\Big)$$
$$= \exp\Big(\tfrac{2\pi i}{N}(\alpha\mu' - \beta\lambda')\Big) \cdot f_{\alpha,\beta}\Big(A\tau, \tfrac{x}{c\tau+d}\Big). \qquad\square$$

7 N-division points and $\Gamma_1(N)\backslash\mathfrak{h}$

7.1 Almost-Chebyshev polynomials and elliptic functions

In the preceding paragraph we have seen that to a given elliptic curve $E = \mathbb{C}/L$ with a fixed non-zero N-division point $Q = \omega + L$ (i.e. $N \cdot \omega \in L$ and $\omega \notin L$) there exists a theta function $f(x)$ for which:

— $\operatorname{div}(f) = (0) - (Q)$,
— $f^N(x)$ is doubly periodic with respect to L,
— $f(x) = x + O(x^2)$ for $x \to 0$,

and that f is uniquely determined by these properties.

From the definition it follows immediately that

$$\operatorname{div}\big(f(x) \cdot f(-x + \omega)\big) = 0$$

and so

$$f(x) \cdot f(-x + \omega) \equiv c^{-2} \neq 0$$

for a suitable constant c. Thus $(cf)^N + (cf)^{-N}$ induces a function on $X = (\mathbb{C}/L)/\langle\tau\rangle$, where $\tau : x \mapsto \omega - x$. The same holds true for f'/f since its pole divisor $-(0) - (\omega)$ is invariant under τ, and its residues are $+1$ and -1, respectively, whence $f'/f =$

$f'/f \circ \tau + \text{const.}$, and const. $= 0$, as follows by setting $x = w/2$ in this equation. Moreover, this also shows that f'/f, as function on X, has exactly one pole which is of order 1, i.e. f'/f defines a biholomorphic map between X and $P_1(\mathbb{C})$ and thus generates the function field of X. Thus there exists a rational function P so that

$$P(f'/f) = f^{-N} + c^{2N} f^N. \tag{1}$$

Clearly, P is uniquely determined by this identity. Since $f^{-N} + c^{2N} f^N$, as a function on X, has a pole only at the same point as f'/f, it follows that P is even a polynomial.

As Hirzebruch observed, (1) implies that P is a normalized almost-Chebyshev polynomial of degree N, i.e.

- $\quad P(x) = x^N + \cdots,$
- $\quad P'(x) = N \cdot x \cdot (x - \xi_1) \cdots (x - \xi_{N-2}),$
- $\quad P(\xi_i) = \pm P(\xi_j)$ for all i and j.

Its values at the critical points ξ_i are $\pm 2c^N$.

Theorem 7.1: *Let $\mathcal{P}_N$ be the set of normalized almost-Chebyshev polynomials of degree N. For $N \geq 3$ the correspondence $(\mathbb{C}/L, Q) \mapsto P$ induces an injection*

$$\{(\mathbb{C}/L, Q) \mid N \cdot Q = 0, \, Q \neq 0\}/\text{isom} \to \mathcal{P}_N/ \sim . \tag{2}$$

Remark: Here, two pairs $(\mathbb{C}/L, Q)$ and $(\mathbb{C}/\widetilde{L}, \widetilde{Q})$ of elliptic curves with distinguished N-division points are defined to be isomorphic if and only if there is a biholomorphic mapping $I : \mathbb{C}/L \to \mathbb{C}/\widetilde{L}$ which carries O to O and Q to $\widetilde{Q}$. The equivalence relation on the set of almost Chebyshev polynomials is defined by

$$F(x) \sim G(x) \;\Leftrightarrow\; F(x) = \lambda^{-N} G(\lambda x) \text{ for some } \lambda \in \mathbb{C}^*.$$

For the proof of Theorem 7.1 we need the following

Proposition 7.2: *Let $I : \mathbb{C}/L \to \mathbb{C}/\widetilde{L}$ be holomorphic with $I(0) = 0$. Then there exists a constant $\lambda \in \mathbb{C}$ so that the following diagram commutes:*

$$
\begin{array}{ccc}
\mathbb{C} & \xrightarrow{\cdot \lambda} & \mathbb{C} \\
\downarrow{\scriptstyle \pi} & & \downarrow{\scriptstyle \widetilde{\pi}} \\
\mathbb{C}/L & \xrightarrow{\;I\;} & \mathbb{C}/\widetilde{L}.
\end{array}
$$

Proof: Since $(\mathbb{C}, \widetilde{\pi})$ is the universal covering of $\mathbb{C}/\widetilde{L}$, we deduce from general theorems of topology the existence of a holomorphic mapping $\widehat{I}$ so that the following diagram is commutative:

$$
\begin{array}{ccc}
\mathbb{C} & \xrightarrow{\widehat{I}} & \mathbb{C} \\
\downarrow{\scriptstyle \pi} & & \downarrow{\scriptstyle \widetilde{\pi}} \\
\mathbb{C}/L & \xrightarrow{\;I\;} & \mathbb{C}/\widetilde{L}.
\end{array}
$$

Now let $\gamma \in L$. Then $x \mapsto \widehat{I}(x + \gamma) - \widehat{I}(x)$ only takes values in $\widetilde{L}$ and is therefore constant, i.e. $\frac{d}{dx}\widehat{I}(x + \gamma) = \frac{d}{dx}\widehat{I}(x)$ for all $\gamma \in L$. Hence $\frac{d}{dx}\widehat{I}$ is holomorphic and elliptic with respect to L, thus is constant. Hence $\widehat{I}(x) = \lambda x + \mu$ for suitable $\lambda, \mu \in \mathbb{C}$, where we have $\mu = \widehat{I}(0) \in \widetilde{L}$, and can therefore without restriction assume that $\mu = 0$. $\quad\square$

Conversely, multiplication with a $\lambda \in \mathbb{C}$ for which $\lambda L \subset \widetilde{L}$ obviously induces a holomorphic mapping $\mathbb{C}/L \rightarrow \mathbb{C}/\widetilde{L}$. Therefore one immediately obtains from the proposition the

Corollary 7.3: *Two pairs* $(\mathbb{C}/L, Q)$ *and* $(\mathbb{C}/\widetilde{L}, \widetilde{Q})$ *are isomorphic if and only if there exists a* $\lambda \in \mathbb{C}$ *with* $\lambda L = \widetilde{L}$ *and* $\lambda Q = \widetilde{Q}$. $\quad\square$

Proof of the theorem: We show first that a mapping (2) is indeed induced. Assume that one starts with two isomorphic pairs $(\mathbb{C}/L, Q)$ and $(\mathbb{C}/\widetilde{L}, \widetilde{Q})$ with $\lambda L = \widetilde{L}$ and $\lambda Q = \widetilde{Q}$, and has obtained the equations $P(f'/f) = f^{-N} + c^{2N} \cdot f^N$ and $\widetilde{P}(\widetilde{f}'/\widetilde{f}) = \widetilde{f}^{-N} + \widetilde{c}^{2N} \cdot \widetilde{f}^N$. Because of the uniqueness of $\widetilde{f}$, we first obtain $\widetilde{f} = \lambda \cdot f(x/\lambda)$ and then

$$\widetilde{c}^{-2} = \widetilde{f}(x) \cdot \widetilde{f}(-x + \widetilde{\omega}) = \lambda^2 f(x/\lambda) \cdot f(-x/\lambda + \omega) = \lambda^2 c^{-2}.$$

Therefore

$$\widetilde{P}\Big(\lambda^{-1}\frac{f'(x/\lambda)}{f(x/\lambda)}\Big) = \widetilde{P}\Big(\frac{\widetilde{f}'(x)}{\widetilde{f}(x)}\Big)$$
$$= \frac{1}{\lambda^N f(x/\lambda)^N} + \lambda^{-2N} c^{2N} \lambda^N f(x/\lambda)^N$$
$$= \lambda^{-N} P\Big(\frac{f'(x/\lambda)}{f(x/\lambda)}\Big),$$

whence $P(x) = \lambda^N \widetilde{P}(\lambda^{-1}x)$.

We now show the injectivity of the mapping (2). So let $P(x)$ be given: then c^{2N} is uniquely determined, since the critical values of P at points unequal to zero are $\pm 2c^N$. (Note that this argument does not hold true for $N = 2$.) Thus P uniquely determines the differential equation $P(f'/f) = f^{-N} + c^{2N} f^N$, which in turn uniquely determines f. For the latter simply solve for the Laurent expansion of f'/f around 0. Finally, f uniquely determines L and Q, its zeroes and poles. This completes the proof of the theorem. $\quad\square$

Proposition 7.4: *The correspondence* $\tau \mapsto (\mathbb{C}/2\pi i(\mathbb{Z}\tau + \mathbb{Z}), 2\pi i/N)$ *induces a bijection*

$$\Gamma_1(N)\backslash\mathfrak{h} \rightarrow \{(\mathbb{C}/L, Q) \mid Q \text{ is a primitive } N\text{-division point}\}/\text{isom}. \tag{3}$$

Remark:

1) Recall that an N-division point is called primitive if N is its precise order in $\mathbb{C}/L$.

2) One can obviously combine the bijections of (3) to obtain a bijection

$$\bigcup_{\substack{N'|N \\ N' \neq 1}} \Gamma_1(N')\backslash\mathfrak{h} \to \{(\mathbb{C}/L, Q) \mid N \cdot Q = 0, Q \neq 0\}/\text{isom}.$$

Proof of the proposition: The mapping is well-defined, i.e. for $A = \left(\begin{smallmatrix} a & b \\ c & d \end{smallmatrix}\right) \in \Gamma_1(N)$ and $\tau' = A\tau$ one obtains the same image:

$$\begin{aligned}
(\mathbb{C}/2\pi i(\mathbb{Z}\tau' + \mathbb{Z}), 2\pi i/N) &= (\mathbb{C}/2\pi i(\mathbb{Z}\tfrac{a\tau+b}{c\tau+d} + \mathbb{Z}), 2\pi i/N) \\
&\cong (\mathbb{C}/2\pi i(\mathbb{Z}(a\tau + b) + \mathbb{Z}(c\tau + d)), 2\pi i(c\tau + d)/N) \\
&= (\mathbb{C}/2\pi i(\mathbb{Z}\tau + \mathbb{Z}), 2\pi i/N),
\end{aligned}$$

where we used that $\frac{c\tau+d}{N} \equiv \frac{1}{N}$ modulo $\mathbb{Z}\tau + \mathbb{Z}$ for $A \in \Gamma_1(N)$.
On surjectivity: In order to construct a preimage of $(\mathbb{C}/L, Q)$ we use the following

Lemma 7.5: *For a primitive N-division point ω, there exists $\overline{\omega} \in L$ and $\omega' \equiv \omega \ (L)$ so that $\overline{\omega}$ and $N \cdot \omega'$ are a $\mathbb{Z}$-basis of L.*

Proof of the lemma: Let $L = \mathbb{Z}\omega_2 + \mathbb{Z}\omega_1$ and $\omega = \frac{\alpha\omega_1+\beta\omega_2}{N}$ with $\alpha, \beta \in \mathbb{Z}$ and $\mathrm{Im}\,(\omega_2/\omega_1) > 0$. Since ω is primitive, we have $(\alpha, \beta, N) = 1$. Now use the easily checked

Lemma 7.6: *For numbers α, β, N with $(\alpha, \beta, N) = 1$, there exist α' and β' with $(\alpha', \beta') \equiv (\alpha, \beta)\ (N)$ and $(\alpha', \beta') = 1$.*

We therefore define ω' by $\omega' := \frac{\alpha'\omega_1+\beta'\omega_2}{N}$ and choose $\gamma, \delta \in \mathbb{Z}$ with $\alpha'\delta - \beta'\gamma = 1$. Then

$$\begin{pmatrix} N\omega' \\ \overline{\omega} \end{pmatrix} = \begin{pmatrix} \alpha' & \beta' \\ \gamma & \delta \end{pmatrix} \begin{pmatrix} \omega_1 \\ \omega_2 \end{pmatrix}$$

defines a basis $N\omega'$ and $\overline{\omega}$ of L with $\mathrm{Im}\,(\overline{\omega}/N\omega') > 0$.
Using Lemma 7.5, we write $(\mathbb{C}/L, Q) = (\mathbb{C}/L, \omega + L)$ as $(\mathbb{C}/(\mathbb{Z}\overline{\omega} + \mathbb{Z}N\omega'), \omega' + L)$. The latter is obviously isomorphic to $(\mathbb{C}/2\pi i(\mathbb{Z}\frac{\overline{\omega}}{N\omega'} + \mathbb{Z}), \frac{2\pi i}{N})$. A preimage is therefore $\tau = \overline{\omega}/(N\omega')$.
On injectivity: Assume that

$$(\mathbb{C}/2\pi i(\mathbb{Z}\widetilde{\tau} + \mathbb{Z}), 2\pi i/N) \cong (\mathbb{C}/2\pi i(\mathbb{Z}\tau + \mathbb{Z}), 2\pi i/N).$$

Then there is a matrix $A = \left(\begin{smallmatrix} a & b \\ c & d \end{smallmatrix}\right) \in \mathrm{SL}_2(\mathbb{Z})$ so that $\widetilde{\tau} = \frac{a\tau+b}{c\tau+d}$. Hence

$$\begin{aligned}
(\mathbb{C}/2\pi i(\mathbb{Z}\widetilde{\tau} + \mathbb{Z}), 2\pi i/N) &\cong (\mathbb{C}/2\pi i(\mathbb{Z}(a\tau + b) + \mathbb{Z}(c\tau + d)), 2\pi i(c\tau + d)/N) \\
&\cong (\mathbb{C}/2\pi i(\mathbb{Z}\tau + \mathbb{Z}), 2\pi i(c\tau + d)/N).
\end{aligned}$$

The latter is by assumption isomorphic to $(\mathbb{C}/(\mathbb{Z}\tau + \mathbb{Z}), 2\pi i/N)$. Therefore we must have $(c\tau + d)/N \equiv 1/N$ modulo $\mathbb{Z}\tau + \mathbb{Z}$, and so $A \in \Gamma_1(N)$. $\square$

The composite of the maps (2) and (3) gives a map

$$\Gamma_1(N)\backslash\mathfrak{h} \to \mathcal{P}_N/\sim\ .$$

This map is induced by the correspondence which associates to each $\tau \in \mathfrak{h}$ the polynomial $P_\tau(x)$ determined by

$$P_\tau\left(\frac{\frac{\partial}{\partial x}f(\tau,x)}{f(\tau,x)}\right) = f(\tau,x)^{-N} + c^{2N}f(\tau,x)^N,$$

where $f(\tau,x) = f_{(0,1)}(\tau,x)$ in the notation of Theorem 6.3 and

$$c^{-2} = f(\tau,x) \cdot f(\tau, -x + 2\pi i/N).$$

This map is holomorphic in the sense that the coefficients of $P_\tau(x)$ are holomorphic. More precisely, one has

Lemma 7.7: *The coefficient* $a_i(\tau)$ *of* $P_\tau(X) = X^N + a_1(\tau)X^{N-1} + \cdots + a_N(\tau)$ *is a modular form of weight* i *on* $\Gamma_1(N)$.

Proof: Let $A_i(L,Q)$ denote the coefficients of the polynomial $P(x)$ associated to the pair $(\mathbb{C}/L, Q)$. Clearly $a_i(\tau) = A_i(2\pi i(\mathbb{Z}\tau + \mathbb{Z}), 2\pi i/N)$. By Theorem 7.1 we have $A_i(\lambda L, \lambda Q) = \lambda^{-i}A_i(L,Q)$. The coefficient $A_i(L,Q)$ is therefore a homogeneous lattice function of degree $-i$. From the discussion at the end of part 3 of §2 we deduce that a_i transforms like a modular form of weight i on $\Gamma_1(N)$.

The regularity of the $a_i(\tau)$ follows from the fact that they are polynomials in certain modular forms on $\Gamma_1(N)$. Indeed, by formulas (6)–(7) of §6 we have

$$f(\tau,x) = x \cdot \exp\Big(2\sum_{k=1}^\infty f_k(\tau)\frac{x^k}{k!} - 2\sum_{\substack{k=1 \\ k\equiv 0\,(2)}}^\infty G_k(\tau)\frac{x^k}{k!}\Big)$$

with f_k (k odd) and $f_k - G_k$ (k even) in $M_k(\Gamma_1(N))$. The coefficients of the Laurent expansion of $f(\tau,x)$ and $\big(\frac{\partial}{\partial x}f(\tau,x)\big)/f(\tau,x)$ around $x = 0$ are therefore polynomials in the f_k (k odd) and $f_k - G_k$ (k even). But then the coefficients of $P_\tau(x)$ are also polynomials in these functions as follows easily from the differential equation for $f(\tau,x)$ and a simple induction argument. $\square$

7.2 The genus of the modular curve $X_1(N)$

In the following let $X_1(N)$ denote the Riemann surface

$$X_1(N) := \overline{\Gamma_1(N)\backslash\mathfrak{h}} = \Gamma_1(N)\backslash\mathfrak{h} \cup \Gamma_1(N)\backslash P_1(\mathbb{Q}).$$

For small values of N one has for the genus $g(N)$ of $X_1(N)$:

N	$1-10$	11	12	13	14	15	16	17	18	≥ 19	≥ 31
$g(N)$	0	1	0	2	1	1	2	5	2	> 2	> 10

Genus $g(N)$ of $X_1(N)$

In this section we derive a formula for the genus $g(N)$. For doing this we have to describe first of all the complex structure of $X_1(N)$, or, more generally, the complex structure of $X(\Gamma) := \Gamma\backslash(\mathfrak{h} \cup P_1(\mathbb{Q}))$ for any subgroup of finite index in $\mathrm{SL}_2(\mathbb{Z})$ (for details compare [Sh71]). So let $\pi : \mathfrak{h} \cup P_1(\mathbb{Q}) \to X(\Gamma)$ be the natural projection. For each $\tau_0 \in \mathfrak{h} \cup P_1(\mathbb{Q})$ we define the following function:

$$t_{\tau_0}(\tau) = \begin{cases} \tau - \tau_0, & \text{if } \tau_0 \in \mathfrak{h}, \mathrm{P\Gamma}_{\tau_0} = \{\mathrm{id}\}, \\ \left(\frac{\tau-\tau_0}{\tau-\overline{\tau_0}}\right)^\nu, & \text{if } \tau_0 \in \mathfrak{h}, |\mathrm{P\Gamma}_{\tau_0}| = \nu > 1, \\ e^{2\pi i A\tau/\nu}, & \text{for } A\tau_0 = \infty, A \in \mathrm{SL}_2(\mathbb{Z}), \nu = [\mathrm{PSL}_2(\mathbb{Z})_{\tau_0} : \mathrm{P\Gamma}_{\tau_0}]. \end{cases}$$

We make $\mathfrak{h} \cup P_1(\mathbb{Q})$ into a topological space by calling a subset M open if for each point $\tau \in \mathfrak{h} \cap M$ an open neighborhood of τ in $\mathfrak{h}$ is contained in M, and for each $\tau \in P_1(\mathbb{Q}) \cap M$ the interior of an open disc touching τ and lying in $\mathfrak{h}$ is contained in M (if $\tau = \infty$, then by definition such a circle is of the form $\{\tau' \in \mathfrak{h} \mid \mathrm{Im}\,(\tau') > \varepsilon > 0\}$). The set $X(\Gamma)$ is given the quotient topology. Then for each $p_0 = \pi(\tau_0) \in X(\Gamma)$ there exist open neighborhoods U of p_0 and V of τ_0 so that $\widetilde{t}(p) := t_{\tau_0}\big((\pi|_V)^{-1}(p)\big)$ defines a homeomorphism of U with a neighborhood of 0. We call $(U,\widetilde{t})$ a chart at p_0. One verifies easily that a complex structure on $X(\Gamma)$ is thereby determined.

The natural mapping

$$\Xi : X_1(N) = \overline{\Gamma_1(N)\backslash\mathfrak{h}} \to \overline{\mathrm{SL}_2(\mathbb{Z})\backslash\mathfrak{h}} \tag{4}$$

is holomorphic and obviously of degree $\mu = [\mathrm{PSL}_2(\mathbb{Z}) : \mathrm{P\Gamma}_1(N)]$. We leave it as an exercise (but compare formula (2) from §1) to verify that

$$\mu = \begin{cases} 3, & \text{if } N = 2, \\ \frac{1}{2} \cdot \varphi(N) \cdot \psi(N), & \text{otherwise}, \end{cases}$$

where $\varphi(N)$ denotes the Euler φ-function

$$\varphi(N) = \sharp\{a \in \mathbb{Z}/N\mathbb{Z} \mid (a, N) = 1\} = N \prod_{p|N}(1 - p^{-1})$$

and

$$\psi(N) = \sharp P_1(\mathbb{Z}/N\mathbb{Z}) = N \prod_{p|N}(1 + p^{-1}).$$

To obtain explicit formulas for $g(N)$ we shall apply the Riemann-Hurwitz formula to the covering (4). For this we need information about possible fixed points and the number of cusps of $\Gamma_1(N)$ since these points are possible ramification points of the map (4).

Lemma 7.8: *Any fixed point τ of $\Gamma \subset \mathrm{SL}_2(\mathbb{Z})$ is equivalent under $\mathrm{SL}_2(\mathbb{Z})$ to*

1) i, *and* $\mathrm{P}\Gamma_\tau$ *is conjugate to* $\langle\left(\begin{smallmatrix} 0 & -1 \\ 1 & 0 \end{smallmatrix}\right)\rangle$ *in* $\mathrm{PSL}_2(\mathbb{Z})$ *and has order* 2, *or*

2) $\rho = e^{2\pi i/3}$, *and* $\mathrm{P}\Gamma_\tau$ *is conjugate to* $\langle\left(\begin{smallmatrix} 0 & -1 \\ 1 & 1 \end{smallmatrix}\right)\rangle$ *in* $\mathrm{PSL}_2(\mathbb{Z})$ *and has order 3.* ▱

We leave the proof of Lemma 7.8 to the reader (or else cf. [Se70]). Note that the lemma implies in particular that any subgroup $\Gamma \subset \mathrm{SL}_2(\mathbb{Z})$ of finite index has, modulo Γ, only finitely many inequivalent fixed points in $\mathfrak{h}$. Moreover it is clear that for any $\tau \in \mathfrak{h}$ one has $|\mathrm{P}\Gamma_\tau| \in \{1, 2, 3\}$.

Lemma 7.9: *Let Γ be a subgroup of finite index in $\mathrm{SL}_2(\mathbb{Z})$, and let $g(\Gamma)$ denote the genus of $X(\Gamma) = \Gamma\backslash(\mathfrak{h} \cup P_1(\mathbb{Q}))$. Let $\mu = [\mathrm{PSL}_2(\mathbb{Z}) : \mathrm{P}\Gamma]$ and let ν_2 and ν_3 denote the number of inequivalent fixed points modulo Γ, which are equivalent modulo $\mathrm{SL}_2(\mathbb{Z})$ to i, resp. $\rho = e^{2\pi i/3}$. Further, let ν_∞ be the number of cusps $\Gamma\backslash P_1(\mathbb{Q})$ of Γ. We then have:*

$$g(\Gamma) = 1 + \frac{\mu}{12} - \frac{\nu_2}{4} - \frac{\nu_3}{3} - \frac{\nu_\infty}{2}.$$

Proof: Let $e_1, \ldots, e_t$ be the branching orders of the natural map $\Xi : X(\Gamma) \to X(1)$ for the preimages of ρ. Since $e_j \mid 3$, e_j is either 1 or 3 and so $\nu_3 = \sharp\{j \mid e_j = 1\}$. Further, $\mu = \sum_{j=1}^{t} e_j$ (if $A_i \in \mathrm{SL}_2(\mathbb{Z})$ such that $A_1\rho, \ldots, A_t\rho$ are representatives for the preimages of ρ then $\bigcup_{i=1,\ldots,t} A_i\Gamma_\rho$ gives a complete set of representatives for $\mathrm{P}\Gamma\backslash\mathrm{PSL}_2(\mathbb{Z})$) and therefore

$$\sum_{j=1}^{t} (e_j - 1) = \frac{2}{3}(\mu - \nu_3).$$

For i and ∞ we have analogously $\sum_{\Xi(P)=i} (e_P - 1) = \frac{1}{2}(\mu - \nu_2)$ and $\sum_{\Xi(P)=\infty} (e_P - 1) = \frac{1}{2}(\mu - \nu_\infty)$. The Riemann-Hurwitz formula therefore says

$$2g(\Gamma) - 2 = \mu(2g(1) - 2) + \sum_{P \in X(\Gamma)} (e_P - 1)$$

$$= -2\mu + \frac{2}{3}(\mu - \nu_3) + \frac{1}{2}(\mu - \nu_2) + (\mu - \nu_\infty),$$

where we have used $g(1) = 0$ (since $j(\tau) = E_4^3(\tau)/\Delta(\tau)$ has as sole singularity a pole of order one and so defines an isomorphism $X_1(1) \to P_1(\mathbb{C})$). Hence the assertion follows. ▱

According to the preceeding lemma we have to look for possible fixed points of $\Gamma_1(N)$:

Lemma 7.10: *Let $P \in \mathfrak{h}$ be a fixed point of $\Gamma_1(N)$. Then either*

$$N = 1 \quad \text{and} \quad P \equiv i \, (\Gamma_1(1)) \quad \text{or} \quad P \equiv \rho \, (\Gamma_1(1)),$$

or

$$N = 2 \quad \text{and} \quad P \equiv \frac{1}{2}(i+1) \, (\Gamma_1(2)),$$

or

$$N = 3 \quad \text{and} \quad P \equiv \frac{1}{6}\left(3 + \sqrt{-3}\right) \, (\Gamma_1(3)).$$

In particular, for $N \geq 4$ the group $\Gamma_1(N)$ has no fixed point.

Proof: For $N = 1$ we know this from Lemma 7.8. From this lemma it is also easily deduced that $\mathrm{tr}\,(A) \in \{\pm 1, 0\}$ for matices $A \in \Gamma_1(N)_\tau$. Now $A \equiv \left(\begin{smallmatrix} 1 & * \\ 0 & 1 \end{smallmatrix}\right) (N)$ and therefore we must have $\mathrm{tr}\,(A) \equiv 2 \,(N)$. This condition cannot be fulfilled for $N \geq 4$. In case $N = 2$ we must have $\mathrm{tr}\,(A) = 0$, hence each fixed point is equivalent to i and of the form $\frac{a \pm i}{c}$ with $a \equiv 1 \,(2)$ and $c \equiv 0 \,(2)$. By Lemma 7.9 we see, since the degree μ of the covering Ξ is 3 in this case, that the number ν_2 of fixed points is equal to 1 or 3. But since $i\Gamma_1(2)$ is not a fixed point (the stabilizer $\langle \left(\begin{smallmatrix} 0 & -1 \\ 1 & 0 \end{smallmatrix}\right)\rangle$ of i does not lie in $\Gamma_1(2)$, $\nu_2 = 1$ and there remains only one fixed point, namely $\frac{1 \pm i}{2}\Gamma_1(2)$. Analogous considerations show that the fixed points of $\Gamma_1(3)$ are conjugate to ρ, and that likewise there is only one of them. $\qquad\qquad$ ⌑

For the number of cusps we have the following

Lemma 7.11: *The number $\mathrm{cu}\,(N)$ of cusps $\Gamma_1(N)\backslash P_1(\mathbb{Q})$ is given by*

$$\mathrm{cu}\,(1) = 1, \ \mathrm{cu}\,(2) = 2, \ \mathrm{cu}\,(4) = 3,$$
$$\mathrm{cu}\,(N) = \frac{1}{2}\sum_{t \mid N} \varphi(t)\varphi(N/t) \quad \text{otherwise.} \tag{5}$$

Proof: For $N = 1$ this is clear. Now each element of $P_1(\mathbb{Q})$ can be written as a column vector $\left(\begin{smallmatrix} a \\ c \end{smallmatrix}\right) = \left(\begin{smallmatrix} a & * \\ c & * \end{smallmatrix}\right)\left(\begin{smallmatrix} 1 \\ 0 \end{smallmatrix}\right)$ for a matrix in $\mathrm{SL}_2(\mathbb{Z})$. This representation is unique up to identification of $\left(\begin{smallmatrix} a \\ c \end{smallmatrix}\right)$ with $\left(\begin{smallmatrix} -a \\ -c \end{smallmatrix}\right)$. In addition, $\Gamma(N) \subset \mathrm{SL}_2(\mathbb{Z})$ is a normal subgroup and $\mathrm{SL}_2(\mathbb{Z})/\Gamma(N) \cong \mathrm{SL}_2(\mathbb{Z}/N\mathbb{Z})$. Therefore the cusps $\Gamma(N)\backslash P_1(\mathbb{Q})$ can be written uniquely (up to sign) as column vectors reduced modulo N (thereby $(a, c, N) = 1$). Under the larger group $\Gamma_1(N)$, $\left(\begin{smallmatrix} a \\ c \end{smallmatrix}\right)$ and the column vector $\left(\begin{smallmatrix} a+xc \\ c \end{smallmatrix}\right) = \left(\begin{smallmatrix} 1 & x \\ 0 & 1 \end{smallmatrix}\right)\left(\begin{smallmatrix} a \\ c \end{smallmatrix}\right)$ reduced modulo N are identified. One can therefore take as representatives of the cusps, up to sign, those $\bar{c} \in \mathbb{Z}/N\mathbb{Z}$ and $\bar{a} \in \mathbb{Z}/(c, N)\mathbb{Z}$ with $(a, c, N) = 1$.

For fixed $t \mid N$ there are $\varphi(N/t)$ possible $\bar{c} \in \mathbb{Z}/N\mathbb{Z}$ for which $t = (c, N)$. For each of these c there are $\varphi(t)$ possible $\bar{a} \in \mathbb{Z}/t\mathbb{Z}$ with $(a, t) = 1$. Hence there are $\sum_{t \mid N} \varphi(t)\varphi(N/t)$ such pairs. Now the pairs $\left(\begin{smallmatrix} a \\ c \end{smallmatrix}\right)$ and $\left(\begin{smallmatrix} -a \\ -c \end{smallmatrix}\right)$ must still be identified. We therefore halve this number apart from $c = 0$ or $c = \frac{N}{2}$. For $c = 0$, $N \neq 2$ we must also halve this number, since $\left(\begin{smallmatrix} a \\ 0 \end{smallmatrix}\right)$ and $\left(\begin{smallmatrix} -a \\ 0 \end{smallmatrix}\right)$ are identified, although $a \not\equiv -a \,(N)$

(consider $t = N$ and $2a \neq N$, since $(a,t) = 1$). For $c = \frac{N}{2}$ and $N \neq 4$ analogous considerations apply. Formula (5) follows from this for $N \neq 2, 4$. For $N = 2$, resp. 4, one thereby makes two, resp. one identification too many, and must therefore add 1, resp. $1/2$ to the result of (5). ▯

We now have

Theorem 7.12: *The genus $g(N)$ of $X_1(N)$ is 0 for $N \leq 4$, and is otherwise given by the formula*

$$g(N) = 1 + \frac{1}{24}\,\varphi(N)\psi(N) - \frac{1}{4}\cdot\sum_{t\mid N}\varphi(t)\varphi(N/t). \tag{6}▯$$

Remarks:

1) For prime numbers $N = p \geq 5$ the formula (6) simplifies to

$$g(p) = \frac{(p-5)(p-7)}{24}.$$

2) Recall from the remark at the end of §1 that $g(N)$ equals the dimension of $S_2(\Gamma_1(N))$. So the theorem gives us an explicit formula for the dimension of the space of cusp forms of weight 2 on $\Gamma_1(N)$. More generally it is possible to show that for $N \geq 5$ and $k \geq 2$ one has

$$\dim M_k(\Gamma_1(N)) = \frac{\varphi(N)\psi(N)}{24}\,(k-1) + \frac{1}{4}\sum_{t\mid N}\varphi(t)\varphi(N/t).$$

Appendix II: The Dirac operator

by Paul Baum, Pennsylvania State University

1 The solution

Let E, F be two complex vector bundles over X, where X is a compact C^∞-manifold without boundary. Further, let $D : C^\infty(E) \to C^\infty(F)$ be an elliptic differential operator; then the kernel and cokernel of D are finite dimensional vector spaces, and the index of D, $\operatorname{ind}(D) := \dim_{\mathbb{C}} \ker(D) - \dim_{\mathbb{C}} \operatorname{coker}(D)$ is well-defined. The Atiyah-Singer index theorem asserts that this index can be computed by a purely topological formula.

If X is complex analytic, E a holomorphic vector bundle over X, and one chooses Hermitian structures for TX and E, then the twisted differential operators $\bar{\partial}_E$ and $\bar{\partial}_E^*$ are defined. If one considers the differential operator

$$\bar{\partial}_E + \bar{\partial}_E^* : C^\infty\left(\left(\sum_{j=0}^{\infty} \Lambda^{0,2j}T^*X\right) \otimes E\right) \longrightarrow C^\infty\left(\left(\sum_{j=0}^{\infty} \Lambda^{0,2j+1}T^*X\right) \otimes E\right),$$

then one obtains as a special case the Riemann-Roch formula of Hirzebruch:

$$\chi(X, E) = (\operatorname{ch}(E) \cdot \operatorname{td}(TX))[X].$$

An important elliptic operator is the Dirac operator. Consider $\mathbb{R}^n$ with $n = 2r$ or $n = 2r + 1$ and matrices $E_1, \ldots, E_n \in M(2^r \times 2^r, \mathbb{C})$ with the following properties (Clifford relations): Each E_j is a skew-adjoint matrix,

$$E_j^2 = -I \quad \text{for } j = 1, \ldots, n$$
$$\text{and} \quad E_j E_k + E_k E_j = 0 \quad \text{for all } j \neq k.$$

Here and in the following I denotes the identity matrix. For odd n, we require that

$$i^{r+1} E_1 E_2 \cdots E_n = I;$$

for even n the E_j are to be block matrices of the form

$$E_j = \begin{pmatrix} 0 & * \\ * & 0 \end{pmatrix} \quad \text{and} \quad i^r E_1 E_2 \cdots E_n = \begin{pmatrix} I & 0 \\ 0 & -I \end{pmatrix}.$$

Such matrices do indeed exist, and there is a simple

Algorithm for construction: For $n = 1$, $E_1 = (-i)$ is uniquely determined. If n is odd, r also increases by one when n is raised to $n + 1$, and $2^r \times 2^r$-matrices $\widetilde{E}_1, \ldots, \widetilde{E}_{n+1}$ with the desired properties are obtained from the matrices $E_1, \ldots, E_n$ for n as follows:

$$\widetilde{E}_j = \begin{pmatrix} 0 & E_j \\ E_j & 0 \end{pmatrix} \quad \text{for } j = 1, \ldots, n \quad \text{and} \quad \widetilde{E}_{n+1} = \begin{pmatrix} 0 & -I \\ I & 0 \end{pmatrix}.$$

Under passage from even n to $n + 1$, r does not change; the new matrices are determined as follows:

$$\widetilde{E}_j = E_j \quad \text{for } j = 1, \ldots, n \quad \text{and} \quad \widetilde{E}_{n+1} = \begin{pmatrix} -iI & 0 \\ 0 & iI \end{pmatrix}.$$

Remark: All entries of these matrices consist of the numbers 0, ± 1 and $\pm i$. In addition, these matrices are unique in the following sense: To any two systems $E_1, \ldots, E_n$ and $\widetilde{E}_1, \ldots, \widetilde{E}_n$ of matrices which satisfy the above properties, there exists a unitary matrix $U \in U(2^r)$ which conjugates the sets into one another:

$$\{\widetilde{E}_j \mid j = 1, \ldots, n\} = \{U E_j U^{-1} \mid j = 1, \ldots, n\}.$$

Example:

$n = 1$: $\quad E_1 = (-i)$.

$n = 2$: $\quad E_1 = \begin{pmatrix} 0 & -i \\ -i & 0 \end{pmatrix}, \quad E_2 = \begin{pmatrix} 0 & -1 \\ 1 & 0 \end{pmatrix}$.

$n = 3$: $\quad E_1 = \begin{pmatrix} 0 & -i \\ -i & 0 \end{pmatrix}, \quad E_2 = \begin{pmatrix} 0 & -1 \\ 1 & 0 \end{pmatrix}, \quad E_3 = \begin{pmatrix} -i & 0 \\ 0 & i \end{pmatrix}$.

We are now in a position to define the Dirac operator for all $\mathbb{R}^n$.

Definition: *The **Dirac operator** for $\mathbb{R}^n$ is defined by*

$$D := \sum_{j=1}^{n} E_j \frac{\partial}{\partial x_j}.$$

Examples:

$n = 1$: $\quad D = -i \frac{\partial}{\partial x}$.

$n = 2$: $\quad D = -2i \cdot \begin{pmatrix} 0 & -\overline{\partial}^* \\ \overline{\partial} & 0 \end{pmatrix}$ with $\overline{\partial} = \frac{1}{2}\left(\frac{\partial}{\partial x_1} + i \frac{\partial}{\partial x_2}\right), \overline{\partial}^* = \frac{1}{2}\left(-\frac{\partial}{\partial x_1} + i \frac{\partial}{\partial x_2}\right)$.

Remark: The Dirac operator D is formally self-adjoint, that is $\int \langle Df, g \rangle \, dx = \int \langle f, Dg \rangle \, dx$ for C^∞-functions $f, g : \mathbb{R}^n \to \mathbb{C}^{2^r}$ with compact support (here $\langle \cdot, \cdot \rangle$ is the standard scalar product on $\mathbb{C}^{2^r}$). It is described by a Hermitian $2^r \times 2^r$ operator matrix. D is a root of the Laplace operator Δ in the following sense: On the strength of the Clifford relations for the E_j, and since the $\frac{\partial}{\partial x_j}$ commute among themselves and with the E_j, we have

$$D^2 = -\Delta \cdot I \quad \text{with} \quad \Delta = \sum_{j=1}^{n} \frac{\partial^2}{\partial x_j{}^2}.$$

In order to construct the Dirac operator on a C^∞-manifold X, one must place two further assumptions on X. Namely, the Stiefel-Whitney classes $w_j \in H^j(X; \mathbb{Z}/2\mathbb{Z})$ must satisfy:

1) $w_1 = 0$ (thus X is orientable),
2) w_2 lies in the image of the reduction homomorphism $\rho : H^2(X; \mathbb{Z}) \to H^2(X; \mathbb{Z}/2\mathbb{Z})$.

When the conditions 1) and 2) are obeyed M admits a so-called Spin^c-structure. For each fixed Spin^c-structure (e.g., complex manifolds have a canonical Spin^c-structure) one can construct a canonical first Chern class c_1. As an example, for complex manifolds the above assumption holds because $\rho(c_1) = w_2$. The ordinary first Chern class thereby coincides with the one obtained from the canonical Spin^c-structure. We use the identity

$$e^{x/2} \cdot \frac{x/2}{\sinh(x/2)} = \frac{x}{1 - e^{-x}}$$

in the ring of formal power series in order to define a Todd class for a manifold X with a Spin^c-structure:

$$\mathrm{td}\,(TX) := e^{c_1/2} \cdot \hat{A}(TX).$$

In the following paragraph, this Todd class solves the problem of computing the index of the Dirac operator.

2   The problem

Definition: *Let $(V, \langle \cdot, \cdot \rangle)$ be an n-dimensional Euclidean vector space, and let $T(V) = \mathbb{R} \oplus V \oplus V \otimes V \oplus \cdots$ be the tensor algebra of V. The tensors of the form $\langle v, v \rangle \cdot 1 + v \otimes v$ generate a two-sided ideal $I \subset T(V)$. The quotient $\mathrm{Cliff}\,(V, \langle \cdot, \cdot \rangle) = \mathrm{Cliff}\,(V) := T(V)/I$ is the **Clifford algebra** of V (with respect to $\langle \cdot, \cdot \rangle$).*

Notice that $\mathbb{R}$ as well as V is canonically embedded in $\text{Cliff}(V)$. In $\text{Cliff}(V)$ we denote the multiplication induced by "$\otimes$" simply with "$\cdot$". The sole relation between the elements can therefore be written as $v^2 = -\langle v, v \rangle \cdot 1$. For an orthonormal basis $e_1, \ldots, e_n$ with respect to $\langle \cdot, \cdot \rangle$ we have the Clifford relations $e_j^2 = -1$ and $e_j \cdot e_k + e_k \cdot e_j = 0$ for $j \neq k$. The Clifford algebra has the 2^n basis elements $e_1^{\varepsilon_1} \cdots e_n^{\varepsilon_n}$ with $\varepsilon_j \in \{0, 1\}$. There is a canonical $\mathbb{R}$-vector space isomorphism

$$e_1^{\varepsilon_1} \cdots e_n^{\varepsilon_n} \rightarrow e_1^{\varepsilon_1} \wedge \cdots \wedge e_n^{\varepsilon_n}$$

of $\text{Cliff}(V)$ onto the exterior algebra $\Lambda(V)$.

Definition: *For an orientable manifold X of dimension $n = 2r$ or $n = 2r+1$, a **spinor system** is a triple $(\varepsilon, \langle \cdot, \cdot \rangle, F)$, where ε is an orientation of X and $\langle \cdot, \cdot \rangle$ is a Riemannian metric on X. One can therefore construct a bundle $\text{Cliff}_{\mathbb{C}}(X)$ of $\mathbb{C}$-algebras, whose fibres over points $P \in X$ are $\text{Cliff}(T_P X, \langle \cdot, \cdot \rangle_P) \otimes_{\mathbb{R}} \mathbb{C}$. In addition, F is a complex vector bundle satisfying the following conditions:*

1) $\text{rk}_{\mathbb{C}} F = 2^r$,
2) *for all $P \in X$, F_P is a Clifford module, i.e. there is a $\mathbb{C}$-linear operation $\sim : \text{Cliff}_{\mathbb{C}}(T_P X) \rightarrow \text{end}(F_P)$ for which 1 acts on each fibre as the identity,*
3) *a Hermitian structure $(\cdot, \cdot)$ is chosen on F, which defines an invariant metric in the following sense:*
4) *each tangent vector $v \in T_P(X) \subset \text{Cliff}_{\mathbb{C}}(T_P X)$ acts as a skew-Hermitian mapping, $\widetilde{v}^* = -\widetilde{v}$.*

In addition, if n is odd and $e_1, \ldots, e_n$ is a positively oriented orthonormal basis of T_P, the product $i^{r+1} \widetilde{e}_1 \cdots \widetilde{e}_n$ is to act as the identity on F_P. For even n, this additional condition need not hold.

In order to construct a spinor system, one needs precisely the assumptions on X which were given in the previous paragraph: $w_1 = 0$ (orientability) and $w_2 \in \rho(H^2(X; \mathbb{Z}))$. Such a manifold together with a spinor system is called a Spin^c-manifold. The choice of the spinor system therefore determines the choice of a preimage $c_1 \in \rho^{-1}(w_2)$.

Exercise: Show that the condition on w_2 is necessary. Find a closed simply-connected manifold X which is not a Spin^c-manifold.

Remarks:

1) In the complex-analytic case, $F = \sum_{j=0}^{\infty} \Lambda^{0,j} T^*$ defines the canonical spinor system.
2) For even n, let $e_1, \ldots, e_n$ be a positively oriented orthonormal basis of $T_P X$. We have $(i^r e_1 \cdots e_n)^2 = \text{id}$. The spinors F therefore decompose under the action of $i^r e_1 \cdots e_n$ into the eigenspaces F_+ of $+1$ and F_- of -1.
3) Originally, TM had $\text{GL}(n, \mathbb{R})$ as structure group. It can be reduced to $\text{GL}^+(n, \mathbb{R})$ by means of ε, and then to $\text{SO}(n)$ by means of the metric $\langle \cdot, \cdot \rangle$. Finally, it can be lifted to a Lie group $\text{Spin}^c(n)$ lying over $\text{SO}(n)$ with fibre S^1.

Choosing a spinor system is equivalent to choosing such a lifting of the structure group of TM from $\mathrm{GL}(n, \mathbb{R})$ to $\mathrm{Spin}^c(n)$.

Definition: *Let X be an n-dimensional manifold with the spinor system $(\varepsilon, \langle \cdot, \cdot \rangle, F)$. A differential operator D is called a **Dirac operator** if it satisfies the following axioms:*

1) $D : C^\infty(F) \to C^\infty(F)$ *is $\mathbb{C}$-linear.*
2)   *The Leibniz rule holds: we have*

$$D(f \cdot s) = (df) \cdot s + f(Ds)$$

    for sections $s \in C^\infty(F)$ and C^∞-functions $f : X \to \mathbb{C}$. Here the cotangent vector df is considered as a tangent vector via $\langle \cdot, \cdot \rangle$ (and then as an element of the Clifford algebra by the canonical embedding).

3)   *The operator D is formally self-adjoint, i.e. for compactly supported sections $s_1, s_2 \in C_c^\infty\left(F|_{X \setminus \partial X}\right)$ we have:*

$$\int_X (Ds_1, s_2)\, d\mathrm{vol} = \int_X (s_1, Ds_2)\, d\mathrm{vol}.$$

4)   *For even n, D must respect the eigenspace decomposition of F:*

$$D^+ := D|_{F_+} : C^\infty(F_+) \to C^\infty(F_-)$$
$$\text{and} \quad D^- := D|_{F_-} : C^\infty(F_-) \to C^\infty(F_+).$$

Remark: Such operators exist. D is uniquely characterized by these axioms up to terms of lower order (in this case, these lower order terms are vector bundle maps). In particular, the symbol and the index are uniquely determined.

Definition: *Let X be a compact Spin^c-manifold without boundary, and E a Hermitian C^∞-vector bundle on X. Then a **twisted Dirac operator** D_E exists,*

$$D_E : C^\infty(F \otimes E) \to C^\infty(F \otimes E).$$

It is elliptic and satisfies the axioms with F replaced by $F \otimes E$.

In particular, for X even-dimensional, the operator $D_E^+ : C^\infty(F_+ \otimes E) \to C^\infty(F_- \otimes E)$ is defined and the solution of the first paragraph yields for its index

$$\mathrm{ind}\left(D_E^+\right) = (\mathrm{ch}\,(E) \cdot \mathrm{td}\,(TX))[X].$$

Remark: For a Spin^c-manifold, c_1 and therefore $\mathrm{td}\,(TX)$ are determined by the choice of the spinor bundle. A different Spin^c-structure on the underlying manifold X leads to a different Dirac operator (having in general a different index).

For $w_2 = 0$, one can choose $c_1 = 0$; The index of the Dirac operator D for this Spin^c-structure is simply $\mathrm{ind}\,(D) = \hat{A}(X)$. Thus for a spin manifold ($w_2 = 0$), the $\hat{A}(X)$-genus $\hat{A}(X)$ and the twisted $\hat{A}$-genera $\hat{A}(X, E)$ are integral.

Appendix III: Elliptic genera of level N for complex manifolds

by Friedrich Hirzebruch

Corrections and improvements in sections 6 and 7 are due to Thomas Berger following a discussion with Serge Ochanine.

(originally published on pages 37–63 in: K. Bleuler, M. Werner (Eds.): *Differential Geometrical Methods in Theoretical Physics (Como 1987)*. NATO Adv. Sci. Inst. Ser. C: Math. Phys. Sci.; 250. Dordrecht, Kluwer Acad. Publ., 1988)

My lecture at the Como Conference was a survey on the theory of elliptic genera as developed by Ochanine, Landweber, Stong and Witten. A good global reference are the proceedings of the 1986 Princeton Conference [1]. In this contribution to the proceedings of the Como Conference I shall not reproduce my lecture, but rather sketch a theory of elliptic genera of level N for compact complex manifolds which I presented in the last part of my course at the University of Bonn during the Wintersemester 1987/88. For a natural number $N > 1$ the elliptic genus of level N of a compact complex manifold M of dimension d is a modular form of weight d on the group $\Gamma_1(N)$. In the cusps of $\Gamma_1(N)$ the genus degenerates either to $\chi_y(M)/(1+y)^d$ where $-y$ is an N-th root of unity different from 1 or to $\chi(M, K^{k/N})$ where K is the canonical line bundle and $0 < k < N$. Here $\chi_y(M) = \sum_{p=0}^{d} \chi^p(M)y^p$ with $\chi^p(M) = \chi(M; \Omega^p) = \sum_{q=0}^{d} (-1)^q h^{p,q}$ is the χ_y-genus introduced in [13] and $\chi(M, K^{k/N})$ is the genus with respect to the characteristic power series

$$\frac{x}{1-e^{-x}} \cdot e^{-(k/N)\cdot x}$$

which equals the holomorphic Euler number of M with coefficients in the line bundle L^k provided $K = L^N$ (see [13]).

For $N = 2$ the genus is expressible in Pontrjagin numbers and hence defined for an oriented differentiable manifold M. The only possible value of $-y$ is -1 and the genus degenerates in the two cusps to

$$\chi_1(M)/2^d = \operatorname{sign}(M)/2^d \quad (\dim_{\mathbb{R}} M = 2d)$$

or to

$$\chi(M, K^{1/2}) = \hat{A}(M).$$

Only recently I realized that Witten in [19] studied also complex manifolds. His discussion includes the genus studied here, at least if one restricts attention to the cusps with specialization $\chi(M, K^{k/N})$.

In this report I shall also try to give an account of the rigidity theorems for complex manifolds with circle actions which for $N = 2$ are due to Taubes [18] and Witten with a new exposition by Bott [9]. These rigidity theorems hold if the first Chern class of M is divisible by N, i.e. if a holomorphic line bundle L with $L^N = K$ exists.

The results in this paper hold also for manifolds with a stable almost complex structure and for circle actions which preserve this structure. For simplicity we have formulated the results for complex manifolds only.

I would like to thank the students of my course Thomas Berger and Rainer Jung for writing the notes. Many thanks to Nils-Peter Skoruppa who lectured several times in my course when I was away and with whom I had helpful discussions on modular forms. After my course I had intensive discussions with Michael Atiyah on the rigidity theorem in Oxford and also with Don Zagier at the Max-Planck-Institut.

1.   In the following N is a fixed natural number > 1, the "variable" x runs through the complex numbers. $\mathfrak{h}$ is the upper half-plane, $\tau \in \mathfrak{h}$ and $q = e^{2\pi i \tau}$. For a lattice L in $\mathbb{C}$ we consider the elliptic function $g(x)$ with divisor $N \cdot 0 - N \cdot \alpha$ where $\alpha \in \mathbb{C}$ is an N-division point ($\alpha \notin L$, $N\alpha \in L$). The function g is uniquely determined by L and α (regarded as element of $\mathbb{C}/L$) if we demand that the power series of g in the origin begins with x^N. The function $f(x) = g(x)^{1/N}$ is uniquely defined if we request $f(x) = x + \text{higher terms}$. The function f is elliptic with respect to a sublattice L' of L whose index in L equals the order of α as element of $\mathbb{C}/L$. For $\omega \in L$ the function $f(x + \omega)/f(x)$ is constant and equals an N-th root of unity. After multiplying L with a non-vanishing complex number we can assume that

$$L = 2\pi i(\mathbb{Z}\tau + \mathbb{Z}) \quad \text{and}$$

$$0 \neq \alpha = 2\pi i\left(\frac{k}{N}\tau + \frac{l}{N}\right) \quad \text{with} \quad 0 \leq k < N \text{ and } 0 \leq l < N. \tag{1}$$

To write down a product development for $f(x)$ in the case that L and α are as in (1) we introduce the entire function[†]

$$\Upsilon(x) = (1 - e^{-x}) \prod_{n=1}^{\infty} (1 - q^n e^{-x})(1 - q^n e^x)/(1 - q^n)^2 \tag{2}$$

[†]   Note that the function $\Upsilon(x)$ equals $e^{-x/2}\Phi(x)$, where $\Phi(x)$ is the function defined in Appendix I, §5.

which has zeroes of order 1 in the points of L. The function $\Upsilon(x)$ equals the Weierstraß sigma-function for L up to a factor of the form $\exp\left(b_1 x + b_2 x^2\right)$. It can be proved easily that

$$f(x) = e^{\frac{k}{N}x}\Upsilon(x)\Upsilon(-\alpha)/\Upsilon(x-\alpha). \tag{3}$$

Namely, it suffices to check

$$\Upsilon(x + 2\pi i\tau) = -e^{-x}e^{-2\pi i\tau}\Upsilon(x).$$

For this replace in (2) the exponential e^x by λ and then substitute λ by λq to see that the factor $-\lambda^{-1}q^{-1}$ comes out. In fact, we have (for $\zeta = e^{2\pi i/N}$)

$$\begin{aligned}
f(x + 2\pi i) &= \zeta^k f(x) \\
f(x + 2\pi i\tau) &= \zeta^{-l} f(x)
\end{aligned} \tag{4}$$

The function $f(x)$ as belonging to L and α (see (1)) degenerates for $q \to 0$ to a function $f_\infty(x)$.

We have

$$\begin{aligned}
f_\infty(x) &= e^{(k/N)x} \cdot \left(1 - e^{-x}\right) \quad \text{for} \quad k > 0 \\
f_\infty(x) &= \left(1 - e^{-x}\right)(1 - e^{\alpha})/\left(1 - e^{\alpha-x}\right) \quad \text{for} \quad k = 0.
\end{aligned} \tag{5}$$

For reasons which are apparent from the introduction we put $e^\alpha = -y$ for $k = 0$ and have in this case

$$f_\infty(x) = \left(1 - e^{-x}\right)(1 + y)/\left(1 + ye^{-x}\right) \quad \text{with} \quad -y = \zeta^l \neq 1.$$

The involution $x \mapsto -x + \alpha$ interchanges the zeroes and poles of $f(x)$. Therefore,

$$f(x)f(-x+\alpha),$$

which is elliptic for L, is in fact a constant $\neq 0$. We write the constant as c^{-2}. Then c^{2N} depends only on L and the chosen N-division point as point of $\mathbb{C}/L$. If the lattice and α are normalized as in (1), then

$$f(x)f(-x+\alpha) = e^{(k/N)\alpha-\alpha}\Upsilon(-\alpha)^2 = c^{-2}$$

and

$$c^{2N} = \Upsilon(-\alpha)^{-2N}q^{\frac{k(N-k)}{N}} \cdot \zeta^{-kl}. \tag{6}$$

The coefficients of the power series developments of $f(x)/x$, $x/f(x)$ and $x\,f'(x)/f(x)$ determine each other. If one replaces in such a series x by λx, one obtains the corresponding function for the lattice $\lambda^{-1}L$ and the N-division point $\lambda^{-1}\alpha$. Therefore the coefficient of x^r in any of these series as function of the pair L, α with $\alpha \in \mathbb{C}/L$

is homogeneous of degree $-r$. Also c^{2N} is such a function of L and α. It is homogeneous of degree $-2N$ and is related to Dedekind's η-function.

If the pair L, α is chosen as in (1), then the coefficients of $f(x)/x$ are indeed modular forms of weight r on the subgroup consisting of the matrices $\left(\begin{smallmatrix} a & b \\ c & d \end{smallmatrix}\right)$ of $\mathrm{SL}_2(\mathbb{Z})$ which satisfy

$$\left(\begin{smallmatrix} a & b \\ c & d \end{smallmatrix}\right)\left(\begin{smallmatrix} k \\ l \end{smallmatrix}\right) \equiv \left(\begin{smallmatrix} k \\ l \end{smallmatrix}\right) \, (N).$$

Also c^{2N} is a modular form of weight $2N$ on this group. (It still has to be shown that these forms are holomorphic in the cusps. See the next section.)

2.   The classification of pairs L, α where L is a lattice in $\mathbb{C}$ and $\alpha \in \mathbb{C}/L$ with $N\alpha = 0$ (but $\alpha \neq 0$), up to multiplication by some complex number $\lambda \neq 0$, leads to the introduction of the modular group

$$\Gamma_1(N) = \{\left(\begin{smallmatrix} a & b \\ c & d \end{smallmatrix}\right) \in \mathrm{SL}_2(\mathbb{Z}) \mid c \equiv 0 \ (N), a \equiv d \equiv 1 \ (N)\}.$$

If we assume that the N-division point has order N in $\mathbb{C}/L$, then the classes of pairs L, α are in one-to-one correspondence with the points of the modular curve $\Gamma_1(N)\backslash\mathfrak{h}$ where $\left(\begin{smallmatrix} a & b \\ c & d \end{smallmatrix}\right)$ acts on $\mathfrak{h}$ by

$$\tau \longmapsto \frac{a\tau+b}{c\tau+d}.$$

This follows from the fact that each pair L, α is equivalent to a pair of type (1) with $\alpha = 2\pi i/N$ (i.e. $k = 0$, $l = 1$). The coefficients of x^r in $x/f(x)$, $f(x)/x$, $xf'(x)/f(x)$ are modular forms of weight r on $\Gamma_1(N)$. It remains to show that such a coefficient is holomorphic in each cusp of $\Gamma_1(N)\backslash\mathfrak{h}$. Transforming $f(x)$ (taken for the lattice (1) with $\alpha = 2\pi i/N$) to a cusp gives a function $f(x)$ for some $\alpha = 2\pi i\left(\frac{k}{N}\tau + \frac{l}{N}\right)$ and the same lattice[†]. The formulas (2) and (3) show immediately that the q-development of each coefficient of $f(x)$ has only non-negative powers of q, fractional if $k > 0$, but then a suitable root of q is the local uniformizing variable at the cusp.

The coefficients e_r of $xf'(x)/f(x)$ are the Eisenstein series. Their q-developments (for $k = 0$ and $l = 1$) for example can be read off from

$$x\frac{f'(x)}{f(x)} = \sum_{n=0}^{\infty} \frac{xq^n e^{-x}}{1 - q^n e^{-x}} - \sum_{n=1}^{\infty} \frac{xq^n e^{x}}{1 - q^n e^{x}} - \sum_{n=0}^{\infty} \frac{x\zeta q^n e^{-x}}{1 - \zeta q^n e^{-x}} + \sum_{n=1}^{\infty} \frac{x\zeta^{-1} q^n e^{x}}{1 - \zeta^{-1} q^n e^{x}}. \quad (7)$$

Furthermore, c^{2N} (see (6)) is a modular form of weight $2N$. For more detailed formulae concerning these q-developments in the case $N = 2$ see [20]. For $N > 2$ and $N \neq 4$ the number of cusps of $\Gamma_1(N)$ equals

$$\frac{1}{2}\sum_{d\mid N} \varphi(d)\varphi\left(\frac{N}{d}\right),$$

where φ is Euler's function[‡]. Each cusp can be represented by several N-division points α as in (1).

[†]   Cf. Appendix I, Theorem 6.4 (ii).

[‡]   Cf. Appendix I, Lemma 7.11.

3.   Let M_d be a compact complex manifold. The Chern classes c_i of M_d are elements of the $2i$-dimensional cohomology group $H^{2i}(M_d; \mathbb{Z})$. Let c be the total Chern class of M_d split up formally

$$c = \sum_{i=0}^{d} c_i = (1 + x_1)(1 + x_2) \cdots (1 + x_d) \tag{8}$$

where $x_1, x_2, \ldots, x_d$ can be regarded as 2-dimensional cohomology classes in some manifold fibered over M_d (see [13], §13.3). Let $Q(x)$ be a fixed power series in the indeterminate x starting with 1 whose coefficients are in some commutative ring containing $\mathbb{Z}$. Then

$$\varphi_Q(M_d) = Q(x_1)Q(x_2) \cdots Q(x_n)[M_d] \tag{9}$$

is the genus of M_d with respect to the power series Q where in (9) the symmetric expression $Q(x_1)Q(x_2) \cdots Q(x_d)$ is written in terms of the Chern classes in view of (8) and the $2d$-dimensional component of this expression is evaluated on M_d (compare [13], §10.2). We define the elliptic genus $\varphi_N(M_d)$ by using the power series

$$Q(x) = \frac{x}{f(x)} = \frac{x\Upsilon(x - \alpha)}{\Upsilon(x)\Upsilon(-\alpha)} \tag{10}$$

where $\alpha = 2\pi i/N$ and $f(x)$ is taken for the pair L, α with $L = 2\pi i(\mathbb{Z}\tau + \mathbb{Z})$. We put again $\zeta = e^{2\pi i/N}$.

Theorem: *The elliptic genus $\varphi_N(M_d)$ is a modular form of weight d on the group $\Gamma_1(N)$. If one represents a cusp of $\Gamma_1(N)$ by $2\pi i\left(\frac{k}{N}\tau + \frac{l}{N}\right)$ with $0 \le k < N$ and $0 \le l < N$, then the value of $\varphi_N(M_d)$ in this cusp equals*

$$\chi\left(M_d, K^{k/N}\right) \quad if \quad k > 0$$

and

$$\chi_y(M_d)/(1 + y)^d \quad if \quad k = 0 \quad and \quad -y = \zeta^l.$$

The theorem follows from the remarks in section 2 and from (5) by recalling that $\chi\left(M_d, K^{k/N}\right)$ is the genus for the power series

$$\frac{x}{1 - e^{-x}} \cdot e^{-(k/N)x}$$

and $\chi_y(M_d)/(1 + y)^d$ is the genus for the power series

$$\frac{x}{1 - e^{-x}} \left(1 + ye^{-x}\right)/(1 + y),$$

see [13].

A genus can be defined also by a power series $Q(x)$ not beginning with 1 (we assume $Q(0) = a_0 \neq 0$). The definition is done by equation (9) again. Then $a_0^{-1} Q(a_0 x)$ gives the same genus with a normalized power series (i.e. the constant term equals 1). We now define $\widetilde{\varphi}_N(M_d)$ using the power series

$$\widetilde{Q}(x) = \frac{x \Upsilon(x - \alpha)}{\Upsilon(x)}, \quad \alpha = 2\pi i / N. \tag{11}$$

Theorem: *The elliptic genus $\widetilde{\varphi}_N(M_d)$ is a modular function on $\Gamma_1(N)$ if $d \equiv 0 \, (2N)$. We have*

$$\varphi_N(M_d) = \widetilde{\varphi}_N(M_d)(\Upsilon(-\alpha))^{-d} = \widetilde{\varphi}_N(M_d) \cdot c^d.$$

The result follows from the preceding theorem and the considerations in section 1 and 2 which show that $\Upsilon(-\alpha)^{-d} = c^d$ is a modular form on $\Gamma_1(N)$ of weight d.

If d is not divisible by $2N$, then

$$\widetilde{\varphi}_N(M_d)^{2N/(d,2N)}$$

is a modular function (here $(d, 2N)$ is the greatest common divisor of d and $2N$). One simply applies the theorem to the $2N/(d, 2N)$-th power of M_d.

The function $\widetilde{\varphi}_N$ has poles in the cusps represented by (1) with $k > 0$. The order of the pole is given by (6). Let us restrict to the case that N is a prime. For $N = 2$ we have 2 cusps represented by $(k, l) = (0, 1)$ and $(k, l) = (1, 0)$. For N odd, we have $2 \cdot \frac{N-1}{2}$ cusps. There are $\frac{N-1}{2}$ cusps represented by $(k, l) = (0, l)$ and $1 \leq l \leq \frac{N-1}{2}$ and $\frac{N-1}{2}$ cusps represented by $(k, l) = (k, 0)$ and $1 \leq k \leq \frac{N-1}{2}$. In the first kind of cusps the q-development of $\widetilde{\varphi}_N(M_d)$ begins with the constant term $\chi_y(M_d)$ with $y = -\zeta^l$, in the latter case it starts with

$$\chi\left(M_d, K^{k/N}\right) \cdot \widetilde{q}^{-k(N-k)d/2N}$$

where $\widetilde{q}$ is a local uniformizing parameter for this cusp of $\Gamma_1(N)\backslash\mathfrak{h}$. (We have $\widetilde{q}^N = q$ in (6)).

4.   For a complex vector bundle W of rank r the exterior powers $\Lambda^i W$ and the symmetric powers $S^i W$ are well-defined vector bundles. Their Chern classes can be calculated from those of W. If $c_1, \ldots, c_r$ are the Chern classes of W (where c_i is in the $2i$-dimensional cohomology of the base space) and if we write formally

$$c = 1 + c_1 + c_2 + \cdots + c_r = (1 + x_1)(1 + x_2) \cdots (1 + x_r)$$

then the Chern character (in the rational cohomology of the base space) is given by

$$\mathrm{ch}\,(W) = e^{x_1} + e^{x_2} + \cdots + e^{x_r}.$$

Over the rationals c and ch determine each other. For the exterior powers we write with some indeterminate t

$$\Lambda_t(W) = \sum_{i=0}^{r} \Lambda^i W \cdot t^i$$

and for the symmetric powers

$$S_t(W) = \sum_{i=0}^{\infty} S^i W \cdot t^i.$$

Then we have for the Chern character

$$\operatorname{ch}(\Lambda_t W) = \prod_{i=1}^{r}(1 + te^{x_i}), \tag{12}$$

$$\operatorname{ch}(S_t W) = \prod_{i=1}^{r}(1 - te^{x_i})^{-1}. \tag{13}$$

Formula (12) was often used in [13]. Formula (13) is, of course, a special case of the general method to calculate the Chern classes associated to given vector bundles by representations [7].

Following Witten's idea (see [19]) we write the elliptic genus $\widetilde{\varphi}_N(M_d)$, or rather its q-development in the standard cusp, in the form

$$\widetilde{\varphi}_N(M_d) = \sum_{n=0}^{\infty} \chi_y(M_d, R_n) q^n. \tag{14}$$

Here, as before, $-y = \zeta = e^{2\pi i/N}$. Furthermore R_n is a virtual vector bundle associated to the complex tangent bundle of M_d by a virtual representation of $\operatorname{GL}(d, \mathbb{C})$ (with coefficients in $\mathbb{Z}[\zeta]$).

For a vector bundle W the polynomial $\chi_y(M_d, W)$ is defined in [13]. We have, if T is the tangent bundle of M_d,

$$\chi_y(M_d, W) = \sum_{p=0}^{d} \chi(M_d, \Lambda^p T^* \otimes W) y^p.$$

We now can specify the R_n in (14). Let us recall that $\widetilde{\varphi}_N(M_d)$ is the genus belonging to the power series (11)

$$\widetilde{Q}(x) = \frac{x}{1 - e^{-x}} \left(1 + ye^{-x}\right) \prod_{n=1}^{\infty} \frac{1 + yq^n e^{-x}}{1 - q^n e^{-x}} \cdot \frac{1 + y^{-1}q^n e^{x}}{1 - q^n e^{x}} \tag{15}$$

Therefore (by (12) and (13))

$$\sum_{n=0}^{\infty} R_n q^n = \prod_{n=1}^{\infty} \Lambda_{yq^n} T^* \cdot \prod_{n=1}^{\infty} \Lambda_{y^{-1}q^n} T \cdot \prod_{n=1}^{\infty} S_{q^n}(T + T^*) \tag{16}$$

(with $-y = \zeta = e^{2\pi i/N}$).

We have

$$R_0 = 1, \quad R_1 = (1 - \zeta)T^* + (1 - \zeta^{-1})T$$

Modulo the ideal $(1 - \zeta)$ of $\mathbb{Z}[\zeta]$, the elliptic genus $\widetilde{\varphi}_N(M_d)$ equals the Euler-Poincaré number $e(M_d)$.

According to Witten's philosophy (compare also [2] and [3]) if we had a χ_y-operator on the loop space $\mathcal{L}M_d$ of M_d, we could try to calculate (or define) its equivariant χ_y-genus for the natural S^1-action on $\mathcal{L}M_d$ with $q \in S^1$ ($q = e^{2\pi i\tau}$, $\tau \in \mathbb{R}$) by the Atiyah-Bott-Singer ([4], [6]) fixed point theorem (fixed point set M_d (constant loops) in $\mathcal{L}M_d$). The result for the equivariant χ_y-genus $\chi_y(\mathcal{L}M_d, q)$ would be that it is the genus with respect to the power series

$$x \prod_{n=-\infty}^{\infty} \frac{1 + yq^n e^{-x}}{1 - q^n e^{-x}}$$

This does not make sense as a power series in q, but formal manipulations bring $\chi_y(\mathcal{L}M_d, q)$ to the genus $\widetilde{\varphi}_n(M_d)$ belonging to $\widetilde{Q}(x)$ (see (15)) provided $(-y)^d = 1$. Observe that (15) is a meromorphic function in the two variables x and q where $(x, q) \in \mathbb{C}^2$ and $|q| < 1$.

5.   The genus $\widetilde{\varphi}_n(M_d)$ has in the standard cusp a development whose coefficients are integral. They are elements of $\mathbb{Z}[\zeta]$. See formula (14). In the cusp (represented by (1) with $0 < k < N$) this is not so. The coefficients are of the form

$$\chi(M_d, K^{k/N} \otimes W_n),$$

where W_n is a virtual vector bundle associated to the tangent bundle by a virtual representation of $GL(d, \mathbb{C})$ with coefficients in $\mathbb{Z}[\zeta]$. The W_n can be calculated using (3). These coefficients are in general not integral. If, however, the first Chern class c_1 of M_d is divisible by N they are integral. This divisibility condition is equivalent to the existence of a holomorphic complex line bundle with $L^N = K$ and the coefficients

$$\chi(M_d, L^k \otimes W_n)$$

become "Riemann-Roch numbers" [13] which are integral.

Theorem: *If the first Chern class c_1 of a complex manifold M_d is divisible by N, then the coefficients of the q-developments of the genus $\widetilde{\varphi}_N(M_d)$ in all cusps (given by (1)) are integral ($\in \mathbb{Z}[\zeta]$); for the elliptic genus $\varphi_N(M_d)$ the coefficients are integral in a cusp with $k > 0$, in the cusps with $k = 0$ they become integral after multiplication with $(1 - \zeta)^d$.*

6.   Let M_d be a compact complex manifold together with an action of the circle S^1 on M_d by holomorphic maps. We write elements of the circle as $\lambda = e^{2\pi i z}$ where $z \in \mathbb{R}/\mathbb{Z}$. The group S^1 acts on the virtual bundles R_n (see (16)). It also acts on the "cohomology group"

$$H^q(M_d; \Lambda^p T^* \otimes R_n) \tag{17}$$

which is in fact a formal direct sum of cohomology groups $H^q(M_d; \Lambda^p T^* \otimes W)$ with coefficients in $\mathbb{Z}(\zeta)$. Since S^1 acts, we get from (17) (considered equivariantly) a character of S^1, i.e. a finite Laurent series in λ. Taking alternating sums over q in (17) gives us a character

$$\chi(M_d, \Lambda^p T^* \otimes R_n, \lambda)$$

and finally (with $-y = e^{2\pi i/N}$),

$$\chi_y(M_d, R_n, \lambda) = \sum_{p=0}^{d} \chi(M_d, \Lambda^p T^* \otimes R_n, \lambda) y^p$$

and

$$\widetilde{\varphi}_N(M_d, \lambda) = \sum_{n=0}^{\infty} \chi_y(M_d, R_n, \lambda) q^n. \tag{18}$$

It may be more convenient to return to our elliptic genus φ_N with characteristic power series $Q(x)$ (see (10)) and consider it equivariantly

$$\varphi_N(M_d, \lambda) = \widetilde{\varphi}_N(M_d, \lambda) \cdot \Upsilon(-\alpha)^{-d}$$
$$= \sum_{n=0}^{\infty} \chi_y(M_d, S_n, \lambda) q^n \tag{19}$$

where the S_n are virtual bundles (coefficients in $\mathbb{Q}(\zeta)$). We can calculate $\varphi_N(M_d, \lambda)$ using the Atiyah-Bott-Singer fixed point theorem (holomorphic Lefschetz theorem [6], p. 566). Before doing this some remarks concerning the fixed point set $M_d^{S^1}$ of the action are necessary. The set $M_d^{S^1}$ is a smooth submanifold of M_d being a disjoint union of connected submanifolds of various dimensions. For each fixed point p, the circle acts in the tangent space T_p, hence integers $m_1, \ldots, m_d$ are defined such that $\lambda \in S^1$ acts by the diagonal matrix $(\lambda^{m_1}, \lambda^{m_2}, \ldots, \lambda^{m_d})$. For each $r \in \mathbb{Z}$ we consider those m_i which are equal to r. This leads to the eigenspace E_r of T_p. Of course, E_0 is the tangent space in p of the connected component of $M_d^{S^1}$ to which p belongs. The numbers $m_1, \ldots, m_d$ (well defined up to order) depend only on the component of $M_d^{S^1}$. Over each component we have eigenspace bundles, also denotes by E_r.

The characteristic power series of the elliptic genus φ_N is given in (10) in the form $Q(x) = x/f(x)$. For the fixed point theorem we need $1/f(x)$. We put

$$F(x) = 1/f(x) = \frac{\Upsilon(x - \alpha)}{\Upsilon(x)\Upsilon(-\alpha)} \tag{20}$$

We shall now give a formula for $\varphi_N(M_d, \lambda)$ using the holomorphic Lefschetz theorem writing it down in short hand form which will need some explanation:

Let ν be an index for the connected components $\left(M_d^{S^1}\right)_\nu$ of the fixed point set $M_d^{S^1}$. Then

$$\varphi_N(M_d, \lambda) = \sum_\nu \varphi_N(M_d, \lambda)_\nu \tag{21}$$

and

$$\varphi_N(M_d, \lambda)_\nu = (e_0 \cdot F(x_1 + 2\pi i m_1 z) \cdots F(x_d + 2\pi i m_d z))[\left(M_d^{S^1}\right)_\nu] \tag{22}$$

where e_0 is the product of those x_i for which $m_i = 0$. Recall $\lambda = e^{2\pi i z}$. Formulas (21) and (22) have the following meaning. For each component of the fixed point set, e_0 is the Euler class (highest Chern class) of its tangent bundle E_0, the formal roots of the total Chern class of E_0 are the x_i with $m_i = 0$. The x_i with $m_i = r \neq 0$ are the formal roots of the total Chern class of the eigenspace bundle E_r over the component. Thus for E_0 one uses in the above product $xF(x) = Q(x)$ and for E_r ($r \neq 0$) the function $F(x + 2\pi i r z)$ which for $rz \notin \mathbb{Z}\tau + \mathbb{Z}$ has no pole for $x = 0$ (and we use a general z) and hence is a power series in x. Observe that the rotation numbers $(m_1, \ldots, m_d)$ depend on the component and also the meaning of the x_i which are the formal roots of the total Chern class restricted to the component. Then one evaluates the expression in (22) on the component and obtains a power series $\varphi_N(M_d, \lambda)_\nu$ in q

$$\varphi_N(M_d, \lambda)_\nu = \sum_{n=0}^{\infty} c_{\nu,n} q^n \tag{23}$$

with coefficients $c_{\nu,n}$ that are meromorphic functions in $\lambda \in \mathbb{C}^*$ and have possible poles only on S^1. Considering (19) we let $c_n := \chi_y(M_d, S_n, \lambda)$ so that c_n is also a meromorphic function in $\lambda \in \mathbb{C}^*$, in fact it is a finite Laurent series (so holomorphic). From the Lefschetz formula we know that $c_n(\lambda) = \sum_\nu c_{\nu,n}(\lambda)$ for λ any topological generator of S^1 and so we get

$$c_n = \sum_\nu c_{\nu,n} \tag{24}$$

as meromorphic functions.

Looking closer at (22) we see that all $c_{\nu,n}$ have possible poles only at $\lambda^m = 1$ for any rotation number m. Furthermore the local term (23) converges for $|q| < 1$ and $|q|^{1/M} < |\lambda| < |q|^{-1/M}$ ($M = \max_i\{m_i\}$) to a meromorphic function and this function has a meromorphic extension to $\mathbb{C}^*$.

According to (4) our function $F(x)$ has the property

$$F(x + 2\pi i) = F(x), \quad F(x + 2\pi i\tau) = \zeta F(x) \tag{25}$$

where $\zeta = e^{2\pi i/N}$. So the meromorphic extension of $\varphi_N(M_d, \lambda)_\nu$ is an elliptic function for the lattice $\mathbb{Z}N\tau + \mathbb{Z}$. More precisely

$$\begin{aligned}
\varphi_N(M_d, \lambda q)_\nu &= \varphi_N\left(M_d, e^{2\pi i(z+\tau)}\right)_\nu \\
&= \zeta^{m_1 + \cdots + m_d}\varphi_N(M_d, \lambda)_\nu
\end{aligned} \tag{26}$$

The exponent $m_1 + \cdots + m_d$ depends on ν, even the residue class of the exponent modulo N depends on ν in general.

Definition: *The S^1-action on M_d is called N-**balanced** if for the components $\left(M_d^{S^1}\right)_\nu$ of the fixed point set the residue class of $m_1 + \cdots + m_d$ modulo N does not depend on ν. If the action is N-balanced, the common residue class of $m_1 + \cdots + m_d$ is called the **type of the action** and denoted by t.*

From formula (24) we immediately get that $\varphi_N(M_d, \lambda) = \sum_{n=0}^\infty c_n q^n$ also converges on some small annulus containing S^1 (depending on the rotation numbers) to a meromorphic function that has a meromorphic extension to $\mathbb{C}^*$. In this sense formula (21) is an identity between meromorphic functions on $\mathbb{C}^*$. In particular we see that $\varphi_N(M_d, \lambda)$ is an elliptic function for the lattice $\mathbb{Z}N\tau + \mathbb{Z}$. For an N-balanced action (26) finally yields the

Theorem: *For an N-balanced S^1-action of type t on the complex manifold M_d, the equivariant elliptic genus $\varphi_N(M_d, \lambda)$ with $\lambda = e^{2\pi i z}$ is an elliptic function for the lattice $\mathbb{Z} \cdot N\tau + \mathbb{Z}$ which satisfies*

$$\begin{aligned}
\varphi_N(M_d, \lambda q) &= \varphi_N\left(M_d, e^{2\pi i(z+\tau)}\right) \\
&= \zeta^t \varphi_N(M_d, \lambda).
\end{aligned} \tag{27}$$

Remark: Of course, φ_N can be regarded as a function of τ and z. In τ it is a modular form of weight d. In fact, φ_N is a meromorphic Jacobi form on $\Gamma_1(N)$ of weight d and index 0 (see [11]).

7.   We now shall approach the rigidity theorems which under certain conditions state that the finite Laurent series $\chi_y(M_d, R_n, \lambda)$ (see (16) and (18)) do not depend on λ. (Recall $-y = e^{2\pi i/N}$). This rigidity means that the elliptic function $\varphi_N(M_d, \lambda)$ of the preceding theorem is a constant (see (19)), i.e. we have to show that it has no poles. The rigidity results were not included in my course at the University of Bonn. When Michael Atiyah came to Bonn in February 1988 he explained to me Bott's approach [9] and that it is rather close to our old paper [5] and we discussed it in Oxford in March. I did not study Taubes's paper [18] in detail, but rather looked in Bott's report [9]. Then

I carried out the proof for the level N case during my visit in Cambridge (England) in March 1988 as a guest of Robinson college.

Let us consider $\varphi_N(M_d, \lambda)$ and the local terms $\varphi_N(M_d, \lambda)_\nu$ as functions of λ for fixed q with $|q| < 1$. From section 6 we know that they are meromorphic for $\lambda \in \mathbb{C}^*$. From (22) we see that $\varphi_N(M_d, \lambda)_\nu$ has poles only for $mz \in \mathbb{Z}\tau + \mathbb{Z}$ where m is a rotation number $\neq 0$ occurring for the component $\left(M_d^{S^1}\right)_\nu$. Of course, $mz \in \mathbb{Z}\tau + \mathbb{Z}$ means $\lambda^m = q^n$ where $n \in \mathbb{Z}$. So from (21), $\varphi_N(M_d, \lambda)$ has possible poles only for $\lambda^m = q^n$ where m is any rotation number.

Now choose $\lambda_0 \in S^1$ with $\lambda_0^m = 1$ for some rotation number m. If q is fixed with $|q| < 1$, then there is a neighborhood U of λ_0 in $\mathbb{C}^*$ such that no point (λ, q) with $\lambda \in U \setminus \{\lambda_0\}$ lies on one of the curves $\lambda^m = q^n$ and that $\sum_{n=0}^\infty c_{\nu,n} q^n$ as well as $\sum_{n=0}^\infty c_n q^n$ converge on U with $c_{\nu,n}$ and c_n as in section 6. Furthermore $c_n = \sum_\nu c_{\nu,n}$ is holomorphic on U (since it is a finite Laurent series) and the only possible pole of $c_{\nu,n}$ on U is λ_0.

Lemma: *Let U be a domain, $\lambda_0 \in U$. Consider meromorphic functions $b_{\nu,n}$ on U for ν in a finite set S and $n \in \mathbb{N}$ with the following properties:*

   1)   *$b_{\nu,n}$ is holomorphic in $U \setminus \{\lambda_0\}$,*

   2)   *$b_n := \sum_{\nu \in S} b_{\nu,n}$ is holomorphic in λ_0,*

   3)   *$\sum_{n=0}^\infty b_{\nu,n}$ converges compactly in $U \setminus \{\lambda_0\}$ for any $\nu \in S$.*

Then $\sum_{n=0}^\infty b_n$ converges compactly in U and is a holomorphic extension of $\sum_{\nu \in S} \sum_{n=0}^\infty b_{\nu,n}|_{U \setminus \{\lambda_0\}}$.

Proof of the lemma: Let K be a compact set with λ_0 in its interior and let L be a small closed disc around λ_0 with $L \subset \overset{\circ}{K}$. For $0 \leq k \leq m$ we have

$$
|\sum_{n=k}^m (\sum_{\nu \in S} b_{\nu,n})|_K = |\sum_{n=k}^m (\sum_{\nu \in S} b_{\nu,n})|_{\partial K}
$$

$$
\leq |\sum_{n=k}^m (\sum_{\nu \in S} b_{\nu,n})|_{\partial(K \setminus L)}
$$

$$
= |\sum_{n=k}^m (\sum_{\nu \in S} b_{\nu,n})|_{\overline{K \setminus L}}
$$

$$
= |\sum_{\nu \in S} \sum_{n=k}^m b_{\nu,n}|_{\overline{K \setminus L}}
$$

$$
\leq \sum_{\nu \in S} |\sum_{n=k}^m b_{\nu,n}|_{\overline{K \setminus L}}.
$$

The first equation is due to the maximum principle, the second one comes from $\overline{L} \subset \overset{\circ}{K}$, the third one is again the maximum principle, the fourth one follows from holomorphicity of $b_{\nu,n}$ on $\overline{K \setminus L}$. Let now be $\varepsilon > 0$ and put $\varepsilon' := \varepsilon/\sharp S$. There exist n_ν such that for m, $k \geq n_\nu$ we have $|\sum_{n=k}^{m} b_{\nu,n}|_{\overline{K \setminus L}} < \varepsilon'$. Thus we finally see that $|\sum_{n=k}^{m} (\sum_{\nu \in S} b_{\nu,n})|_K < \varepsilon$ for any m, $k \geq n_0 := \max_\nu \{n_\nu\}$.

Using this lemma with $b_{\nu,n} := c_{\nu,n} q^n$ we immediately get that $\varphi_N(M_d, \lambda) = \sum_{n=0}^{\infty} c_n q^n$ has no poles on S^1.

An S^1-action is semi-free (i.e. the fixed point set of any $\lambda \in S^1$, $\lambda \neq 1$, equals $M_d^{S^1}$) if and only if all non-vanishing rotation numbers m equal ± 1. Therefore, for a semi-free action, $M_d^{S^1}$ can have poles only for $\lambda = q^n$ with $n \in \mathbb{Z}$.

Theorem: *For an N-balanced semi-free S^1-action of type t on the complex manifold M_d, the equivariant elliptic genus $\varphi_N(M_d, \lambda)$ does not depend on λ. It equals the elliptic genus $\varphi_N(M_d)$. If $t \not\equiv 0\ (N)$, then $\varphi_N(M_d) = 0$.* (Compare [14] and [15].)

Proof: By the lemma, there is no pole for $\lambda = 1$. Because of (25) there are no poles for $\lambda = q^n$. The vanishing of $\varphi_N(M_d)$ follows also from (25).

8.   Let M_d be a complex manifold with first Chern class $c_1 \in H^2(M_d; \mathbb{Z})$ divisible by N. The importance of this condition was already apparent in section 5. We choose a holomorphic line bundle L with $L^N = K$. Now suppose we have an S^1-action on M_d. Consider the N-fold covering $S^1 \to S^1$ with $\mu \mapsto \lambda = \mu^N$. Then μ acts on M_d and K through λ. This action can be lifted to L. If p is a fixed point of the given S^1-action with rotation numbers $m_1, m_2, \ldots, m_d$, then μ acts in the fibre L_p by $\mu^{-(m_1+\cdots+m_d)}$. However, if $\mu = \zeta = e^{2\pi i/N}$, then it operates trivially on M_d. Therefore the action of ζ in each fibre of L is by multiplication with ζ^{-t}, where t is a residue class modulo N which does not depend on the base point of the fibre. (Assume that M_d is connected.) It follows that the action is N-balanced of type t (see the definition in section 6).

The condition $c_1 \equiv 0\ (N)$ implies a stronger property than N-balanced. Let $G_m \subset S^1$ be the group of m-th roots of unity. The fixed point set of G_m is a submanifold of M_d which includes $M_d^{S^1}$ and is strictly larger if and only if there is a rotation number divisible by m. We denote the fixed point set of G_m by M_d^m. There is the map $S^1 \to S^1$ with $\mu \mapsto \lambda = \mu^N$ which we considered before. Hence any $\mu \in S^1$ with $\mu^{mN} = 1$ operates trivially on M_d^m, however it operates on every fibre L_p ($p \in M_d^m$) by multiplication with some mN-th root of unity which only depends on the connected component of M_d^m which contains p. Since μ acts on L_p (for $p \in M_d^{S^1}$) by $\mu^{-(m_1+\cdots+m_d)}$ where the m_j are the rotation numbers of the action in p, it follows that the residue class of $m_1 + \cdots + m_d$ modulo mN depends only on the connected components of M_d^m and not on the components of $M_d^{S^1}$ contained in them.

Let X be a connected component of M_d^m and $\left(M_d^{S^1}\right)_\nu$ a component of $M_d^{S^1}$ contained in X with rotation numbers $m_1, \ldots, m_d$. Over X the tangent bundle T of M_d splits into vector bundles $\widetilde{E}_k$ where $k = 0, 1, \ldots, m-1$ and the action of G_m in $\widetilde{E}_k$ is by multiplication with λ^k if $\lambda \in G_m$. Of course, $\widetilde{E}_0$ is the tangent bundle of X.

Over $\left(M_d^{S^1}\right)_\nu$ we have

$$\widetilde{E}_k = \sum_{r \equiv k\ (m)} E_r \qquad \text{(see section 6).} \tag{28}$$

We write the rotation numbers in the following form

$$m_i = r_i m + k_i \quad \text{where} \quad k_i = 0, 1, \ldots, m-1. \tag{29}$$

Since the integer $\sum_{i=0}^{d} k_i = \sum_{k=0}^{m-1} k \cdot \operatorname{rk} \widetilde{E}_k$ depends only on X, we see that $m \cdot \sum_{i=0}^{d} r_i$ modulo mN depends only on X. Hence, $\sum_{i=0}^{d} r_i$ modulo N depends only on X and not on the components $\left(M_d^{S^1}\right)_\nu$ contained in it. We put

$$t(m, X) := \sum_{i=0}^{d} r_i \bmod N. \tag{30}$$

Of course, $t(1, M_d)$ is the type t of the action (for connected M_d).

9.　　　Let M_d be a compact complex manifold with $c_1 \equiv 0\ (N)$. We assume that we have an S^1-action and wish to show that the elliptic function $\varphi_N(M_d, \lambda)$ has no poles. Let X be a connected component of M_d^m (see section 8). We define

$$\varphi_N(X, \lambda) := \sum_\nu \varphi_N(M_d, \lambda)_\nu \tag{31}$$

where the summation is over those connected components $\left(M_d^{S^1}\right)_\nu$ which are contained in X (see (21)). This is a shorthand notation. Do not confuse (31) with the elliptic genus of X. Let $\left(M_d^{S^1}\right)_\nu$ have the rotation numbers $m_1, \ldots, m_d$. According to (22) we have

$$\varphi_N(M_d, \lambda)_\nu = (e_0 F(x_1 + 2\pi i m_1 z) \cdots F(x_d + 2\pi i m_d z)) \left[\left(M_d^{S^1}\right)_\nu\right]. \tag{32}$$

Let s be an integer and replace in (32) the variable z by $z + \frac{s}{m}\tau$ (in other words, replace λ by $\lambda \cdot q^{s/m}$), then $\varphi_N\left(M_d, \lambda q^{s/m}\right)_\nu$ is again an elliptic function in z for the lattice $\mathbb{Z} \cdot N\tau + \mathbb{Z}$. It follows from (25), (29) and (30) that

$$\varphi_N\left(M_d, \lambda q^{s/m}\right)_\nu = \zeta^{st(m,X)} \cdot \left(e_0 \prod_{j=1}^{d} F\left(x_j + 2\pi i m_j z + 2\pi i \frac{s k_j}{m}\tau\right)\right)\left[\left(M_d^{S^1}\right)_\nu\right]. \tag{33}$$

If we write down the q-development of the right hand side of (33) (with fractional powers of q) we see that $\varphi_N\left(X, \lambda q^{s/m}\right)$ is of the form

$$\varphi_N\left(X, \lambda q^{s/m}\right) = \sum_{n=0}^{\infty} \chi_y(X, S_n, \lambda) q^{n/m} \tag{34}$$

where the S_n are virtual equivariant bundles constructed from the bundles $\widetilde{E}_k$ over X. For $m = 1$ we come back to (19). The elliptic function (34) has no poles for $|\lambda| = 1$. We use again the lemma in section 7.

10. We are now able to prove the rigidity theorem.

Theorem: *Let M_d be a compact complex manifold with first Chern class $c_1 \in H^2(M_d; \mathbb{Z})$ divisible by N. Suppose an S^1-action on M_d is given. Then the equivariant elliptic genus $\varphi_N(M_d, \lambda)$ does not depend on $\lambda \in S^1$. It equals the elliptic genus $\varphi_N(M_d, \lambda) = (M_d, 1)$. If the type t of the action is $\not\equiv 0 \ (N)$, then $\varphi_N(M_d) \equiv 0$.*

Proof: Let m be a natural number ≥ 1 and $\lambda^{\pm m} = q^n$. Then λ is of the form $\lambda = \lambda_0 q^{s/m}$ where $\lambda_0^m = 1$ and $s = \pm n$. We have

$$\varphi_N\left(M_d, \lambda_0 q^{s/m}\right) = \sum_X \varphi_N\left(X, \lambda_0 q^{s/m}\right)$$

where the summation is over all the connected components of M_d^m. Since the elliptic function (34) has no poles for λ_0, the result follows. The vanishing $\varphi_N(M_d) = 0$ for $t \not\equiv 0 \ (N)$ follows again from (25).

11. We want to point out some applications of the rigidity theorem.

If we develop in a cusp (1) with $k > 0$, we get a different version of the rigidity theorem (compare [19]). In particular, we get that $\chi\left(M_d, L^k, \lambda\right)$ does not depend on λ for $k = 1, \ldots, N-1$, in fact $\chi\left(M_d, L^k\right) = 0$, if the action is non-trivial. This is a well-known result ([12], [15]). For $N = 2$ and $k = 1$ it corresponds to the theorem in [5] on the $\hat{A}$-genus.

The elliptic genus of level N is strictly multiplicative in fibre bundles with a manifold M_d with $c_1(M_d) \equiv 0 \ (N)$ as fibre and a compact connected Lie group G of automorphisms of M_d as structure group (compare [14] and [16]).

This we wish to apply, for example, to the compact irreducible Hermitian symmetric spaces G/U studied in [7], §16. There we gave a formula for the coefficient $\lambda(G/U)$ in

$$c_1(G/U) = \lambda(G/U) \cdot g$$

where g is a positive generator of the infinite cyclic group $H^2(G/U)$.

Take a system $w_1, \ldots, w_d$ of positive complementary roots for G/U (see [7]). Here d is the complex dimension of G/U. The roots $w_1, \ldots, w_d$ are linear forms in $x_1, \ldots, x_l$ where $l = \mathrm{rk}\, U = \mathrm{rk}\, G$, the $x_1, \ldots, x_l$ can be identified with a base of $H^1(T; \mathbb{Z})$ where T is the maximal torus of U. Without proof we state the following result which is equivalent to the strict multiplicativity of the elliptic genus for G/U-bundles.

Theorem: *Let $F(x) = f(x)^{-1}$ be the elliptic function introduced for level N (see section 1). Let $w_1, \ldots, w_d$ be positive complementary roots for the irreducible Hermitian symmetric space G/U. Suppose $\lambda(G/U) \equiv 0 \ (N)$. Then*

$$\sum_{\sigma \in W(G)/W(U)} F(\sigma(w_1))F(\sigma(w_2)) \cdots F(\sigma(w_d)) = \varphi_N(G/U). \tag{35}$$

Here $W(G)$, $W(U)$ are the Weyl groups. (An element $\sigma \in W(U)$ permutes $w_1, \ldots, w_d$. Therefore the sum over the $W(U)$-cosets is well-defined.)

The formula (35) is an identity in the l variables $x_1, \ldots, x_l$. The sum is a constant, i.e. does not depend on these variables anymore.

The rigidity theorem in section 10 also gave a vanishing result. We give an example: Consider the Grassmannian

$$W(m, n) = \mathrm{U}(m + n)/(\mathrm{U}(m) \times \mathrm{U}(n)).$$

We use the notation of [7]. As a system of positive roots of $\mathrm{U}(m + n)$, we take

$$\{-x_i + x_j \mid 1 \le i < j \le m + n\}.$$

The complementary roots w_r are given by $1 \le i \le m$ and $m + 1 \le j \le m + n$. Their sum equals

$$\sum_r w_r = -n \sum_{i=1}^{m} x_i + m \sum_{j=m+1}^{m+n} x_j = -(m + n) \sum_{i=1}^{m} x_i + m \sum_{j=1}^{m+n} x_j. \tag{36}$$

We put $-\sum_{i=1}^{m} x_i = g$ and $\sum_{j=1}^{m+n} x_j = \sigma_1$. Then g becomes the positive generator of $H^2(W(m, n); \mathbb{Z})$ whereas σ_1 vanishes if regarded as element of this cohomology group. Therefore

$$\lambda(W(m, n)) = m + n.$$

We also see from (36) that $W(m, n)$ admits an N-balanced circle action of type m if $m + n \equiv 0 \ (N)$. We obtain

Proposition: *The elliptic genus $\varphi_N(W(m, n))$ vanishes if $m + n \equiv 0 \ (N)$ and $m \not\equiv 0 \ (N)$. For the complex projective spaces $P_n(\mathbb{C}) = W(n, 1)$ we have*

$$\varphi_N(P_n(\mathbb{C})) = 0 \quad if \quad n + 1 \equiv 0 \ (N).$$

12.   The elliptic function f defined in section 1 satisfies a differential equation

$$\left(\frac{f'}{f}\right)^N + a_1\left(\frac{f'}{f}\right)^{N-1} + \cdots + a_{N-1}\left(\frac{f'}{f}\right) + a_N = \frac{1}{f^N} + a_{2N} f^N \tag{37}$$

with $a_{2N} = c^{2N}$ (see section 1) where the a_j are modular forms of weight j for $\Gamma_1(N)$, (if $\alpha = 2\pi i/N$ in (1)). The polynomial

$$P(\xi) = \xi^N + a_1 \xi^{N-1} + \cdots + a_{N-1}\xi + a_N$$

has the following properties:

1) $a_{N-1} = 0$.
2)   If $P'(\xi) = 0$, but $\xi \ne 0$, then $P(\xi)^2 = 4a_{2N}$.

The property 2) implies that the values at the critical points ξ with $\xi \neq 0$ are all equal up to sign. In this case, the polynomial might be called almost-Chebyshev. Theodore J. Rivlin wrote to me that polynomials with essentially such properties occur in the literature under the name Zolotarev-polynomials. Also their relation to elliptic functions is known (see for example [10]). I plan to write a separate paper on these matters. For $N = 2$ the differential equation is of the form

$$f'^2 = 1 - a_2 f^2 + a_4 f^4.$$

very well known for the elliptic genus of level 2 (see [14]).

References

[1] P.S. Landweber (Ed.): *Elliptic Curves and Modular Forms in Algebraic Topology, Proceedings Princeton 1986*. Lecture Notes in Mathematics; 1326. Berlin-Heidelberg, Springer, 1988.

[2] O. Alvarez, T.-P. Killingback, M. Mangano, P. Windey: String Theory and Loop Space Index Theorems, *Comm. Math. Phys.* **111** (1987), pp. 1–10.

[3] O. Alvarez, T.-P. Killingback, M. Mangano, P. Windey: The Dirac-Ramond operator in string theory and loop space index theorems; pp. 189–216 in: H. Hamber (Ed.): *Proc. Conf. on Nonperturbative Methods in field Theory Jan 87, Irvine, Univ. of California*. Proc. Suppl. Sec. *Nucl. Phys. B* **1A** (1987).

[4] M.F. Atiyah, R. Bott: A Lefschetz fixed point formula for elliptic complexes: II. Applications, *Ann. of Math.* **88** (1968), pp. 451–491.

[5] M.F. Atiyah, F. Hirzebruch: Spin-Manifolds and Group Actions; pp. 18–28 in: *Essays on Topology and Related Topics. Memoires dédiés à Georges de Rham*. Berlin-Heidelberg-New York, Springer, 1970.

[6] M.F. Atiyah, I.M. Singer: The index of elliptic operators: III, *Ann. of Math.* **87** (1968), pp. 546–604.

[7] A. Borel, F. Hirzebruch: Characteristic classes and homogeneous spaces, I, *Am. J. Math.* **80** (1958), pp. 458–538.

[8] A. Borel, F. Hirzebruch: Characteristic classes and homogeneous spaces, II, *Am. J. Math.* **81** (1959), pp. 315–382.

[9] R. Bott: On the fixed point formula and the rigidity theorems of Witten lectures at Cargèse 1987; pp. 13–32 in: *Nonperturbative quantum field theory (Cargèse, 1987)*. NATO Adv. Sci. Inst. Ser. B: Phys.; 185. New York, Plenum, 1988.

[10] B.C. Carlson, J. Todd: Zolotarev's first problem — the best approximation by polynomials of degree $\leq n-2$ to $x^n - n\sigma x^{n-1}$ in $[-1,1]$, *Aequationes Math.* **26** (1983), pp. 1–33

[11] M. Eichler and D. Zagier: *The theory of Jacobi forms*. Progress in Mathematics; Vol. 55. Boston-Basel-Stuttgart, Birkhäuser, 1985.

[12] A. Hattori: Spin^c-Structures and S^1-Actions, *Invent. Math.* **48** (1978), pp. 7–31.

[13] F. Hirzebruch: *Topological Methods in Algebraic Geometry* (3rd edition, with appendices by R.L.E. Schwarzenberger and A. Borel). Grundlehren der mathematischen Wissenschaften; 131. Berlin-Heidelberg, Springer, 1966.

[14] P.S. Landweber: Elliptic genera: An introductory overview; pp. 1–10 in [1].

[15] K.H. Mayer: A Remark on Lefschetz Formulae for Modest Vector Bundles, *Math. Ann.* **216** (1975), pp. 143–147.

[16] S. Ochanine: Genres Elliptiques Équivariantes; pp. 107–122 in [1].

[17] G. Segal: Elliptic cohomology, *Séminaire Bourbaki*, 40e année, 1987–88, no. 695 (Février 1988).

[18] C.H. Taubes: S^1 Actions and Elliptic Genera, *Comm. Math. Phys.* **122** (1989), pp. 455–526.

[19] E. Witten: The Index of the Dirac Operator in Loop Space; pp. 161–181 in [1].

[20] D. Zagier: Note on the Landweber-Stong Elliptic Genus; pp. 216–224 in [1].

Appendix IV: Zolotarev polynomials and the modular curve $X_1(N)$

by Rainer Jung, Max-Planck-Institut für Mathematik, Bonn

1   Zolotarev polynomials

In this appendix we want to study the relation between lattices L in $\mathbb{C}$, together with a distinguished N-division point in $\mathbb{C}/L$, and Zolotarev polynomials $P(x)$ of degree N. These polynomials have the following nice property: If we denote the critical points of P by $\xi_0, \ldots, \xi_{N-2}$, then the critical values of P away from ξ_0 are all equal up to sign,

$$P(\xi_i) = \pm P(\xi_j) \quad \text{for } i \neq j \text{ and } 0 < i, j \leq N - 2.$$

We normalize P to $P(x) = x^N + \cdots$ and to $\xi_0 = 0$ and restrict to the case where all ξ_i are distinct. Then we get that $(P'(x)/x)^2$ divides $P(x)^2 - A^2$ with $A = \pm P(\xi_i)$. So there is a polynomial $Q(x) = x^4 + \cdots$ of degree 4, such that the following equation holds:

$$Q \cdot P'^2 = N^2 x^2 (P^2 - A^2).$$

This motivates the following

Definition: *A polynomial $P(x) = x^N + \cdots$ is called a normalized **Zolotarev polynomial** (with critical value $\pm A$), if there exists a polynomial $Q(x) = x^4 + \cdots$ such that*

$$Q \cdot P'^2 = N^2 x^2 (P^2 - A^2). \tag{1}$$

*A Zolotarev polynomial P is called **degenerate** if the discriminant $\mathrm{discr}(Q)$ of Q vanishes (i.e. if Q has multiple zeroes).*

Remarks:

1) For $N > 2$ the polynomial P determines A and Q in (1) uniquely. Conversely P and A are uniquely determined by Q.
2) If $N = 2$, and P is non-degenerate, then (1) reduces to $Q(x) = P(x)^2 - A^2$, so Q resp. the pair (P, A^2) uniquely determine each other.
3) If $N = 1$, equation (1) reduces to $Q(x) = x^2(P(x)^2 - A^2)$, so P is degenerate.

Examples for Zolotarev polynomials are the normalized Chebyshev polynomials $\widetilde{T}_N(x)$ we already used in Chapter 7, section 7.5. They were defined by the property

$$\widetilde{T}_N\left(x + x^{-1}\right) = x^N + x^{-N},$$

and are Zolotarev polynomials for $A = \pm 2$ and $Q(x) = x^2(x-2)(x+2)$. So we see that they are degenerate and furthermore that for the Chebyshev polynomials all the critical values coincide up to sign. All degenerate Zolotarev polynomials can be determined explicitly. By a straightforward calculation (cf. [Ju89]) one gets the

Proposition: *If P is a normalized degenerate Zolotarev polynomial of degree N, then either of the following two cases applies:*

   (i) $P(x) = (x - k\lambda)^k (x + (N-k)\lambda)^{N-k}$ *with* $\lambda \in \mathbb{C}$, $k \in \mathbb{N}$, *and* $0 \le k \le N$. *Equation (1) holds with* $A = 0$ *and* $Q(x) = ((x - k\lambda)(x + (N-k)\lambda))^2$.

   (ii) $P(x) = \mu^N \widetilde{T}_N((x - \lambda)/\mu)$, *with* $\lambda, \mu \in \mathbb{C}$, $\mu \ne 0$. *Equation (1) holds with* $A = \pm 2\mu^N$ *and* $Q(x) = x^2(x - (\lambda - 2\mu))(x - (\lambda + 2\mu))$.

Furthermore $P'(0) = 0$ if and only if $N > 1$ and in the first case $k \ne 0, N$ or $\lambda = 0$ (i.e. $P(x) = x^N$), or in the second case $\frac{\lambda}{\mu} = 2\cos(\frac{l\pi}{N})$ for any l with $0 < l < N$. □

For the following we have to restrict our notion of Zolotarev polynomials:

Definition: *A normalized Zolotarev polynomial P of degree N with critical value $\pm A$ is called **admissible**, if $P'(0) = 0$ and $P(x) \ne x^N$ for $N > 2$, resp. $P(x) \ne x^2$ or $A \ne 0$ for $N = 2$. For $N > 2$ we denote by $\mathcal{P}_N$ (resp. $\mathcal{P}_N^\circ$) the set of admissible normalized (non-degenerate) Zolotarev polynomials P of degree N. For $N = 2$ we denote by $\mathcal{P}_2$ (resp. $\mathcal{P}_2^\circ$) the set of pairs (P, C), such that P is an admissible normalized (non-degenerate) Zolotarev polynomial of degree 2 with critical value $\pm A$, where $A^2 = 4C$. By $\mathcal{Q}_N$ (resp. $\mathcal{Q}_N^\circ$) we denote the set of normalized polynomials of degree 4, fulfilling (1) for some polynomial $P \in \mathcal{P}_N$ (resp. $P \in \mathcal{P}_N^\circ$) for $N > 2$, resp. for some pair $(P, C) \in \mathcal{P}_2$ (resp. $(P, C) \in \mathcal{P}_2^\circ$) for $N = 2$.*

Remark: If P is non-degenerate, then (1) ensures that $P'(0) = 0$ and $P \ne x^N$ for $N > 2$, resp. $(P, A) \ne (x^2, 0)$ for $N = 2$, so P is admissible. Therefore admissability is only a restriction on the set of degenerate Zolotarev polynomials that we already listed above.

For the rest of this appendix we will always use the term Zolotarev polynomial synonymously for admissible Zolotarev polynomial.

From the remarks 1) and 2) above and the list of degenerate Zolotarev polynomials, we get the following

Lemma: *The canonical map $\Omega_N : \mathcal{Q}_N \to \mathcal{P}_N$ mapping a polynomial $Q \in \mathcal{Q}_N$ to the Zolotarev polynomial P defined by (1) (resp. to the pair $(P, A^2/4)$ for $N = 2$) is a bijection. It also induces a bijection from $\mathcal{Q}_N^\circ$ to $\mathcal{P}_N^\circ$.* □

There is also a map connecting Zolotarev polynomials for different degree (we only give the description for $N > 2$, but the construction works analogously for $N = 2$): If Q

is in $\mathcal{Q}_N$ and $P = \Omega_N(Q) \in \mathcal{P}_N$ so that (1) holds for some critical value A, then we have $(A/2)^n \tilde{T}_n(P(x)/(A/2)) \in \mathcal{P}_{nN}$ and equation (1) holds for the same polynomial Q and the critical value $2(A/2)^n$. So we have $\mathcal{Q}_N \subset \mathcal{Q}_{nN}$ and if we denote by ι_n the map

$$\iota_n : \mathcal{P}_N \to \mathcal{P}_{nN}$$
$$P(x) \mapsto (A/2)^n \tilde{T}_n(P(x)/(A/2))$$

we get the following commutative diagram:

$$\begin{array}{ccc} \mathcal{Q}_N & \subset & \mathcal{Q}_{nN} \\ \Omega_N \downarrow & & \downarrow \Omega_{nN} \\ \mathcal{P}_N & \overset{\iota_n}{\to} & \mathcal{P}_{nN} \end{array}$$

Since the polynomial Q does not change when applying the map ι_n to P, the diagram induces a similar one on non-degenerate polynomials. Furthermore, since Ω_N and Ω_{nN} are bijections, the map ι_n is injective and the diagram also gives $\iota_n \circ \iota_m = \iota_{nm}$.

2 Interpretation as an algebraic curve

There exists an action of $\mathbb{C}^*$ on the set of polynomials $F(x)$ of degree r, given by

$$\lambda(F(x)) = F(\lambda x)/\lambda^r \quad \text{for } \lambda \in \mathbb{C}^*.$$

If P is a Zolotarev polynomial of degree N for the critical value A, and Q is the polynomial of degree 4 in equation (1), then $\lambda(P)$ is again a Zolotarev polynomial for the critical value $\lambda^{-N} A$ and satisfies (1) with $\lambda(Q)$. So the action induces an action on $\mathcal{Q}_N$, $\mathcal{Q}_N^\circ$, $\mathcal{P}_N$ and $\mathcal{P}_N^\circ$, where on $\mathcal{P}_2$ (resp. on $\mathcal{P}_2^\circ$) λ acts as $\lambda(P, C) = (\lambda(P), \lambda^{-4} C)$. The action is compatible with the maps Ω_N and ι_n. We denote the induced equivalence relation by " $\sim$ ".

We want to identify the polynomials with their coefficients. Later on we will be interested in equivalence classes of polynomials, so we make the

Definition: *For $n_0, \ldots, n_r \in \mathbb{N}$, the **weighted projective space** $P^{n_0, \ldots, n_r}(\mathbb{C})$ is defined by $(\mathbb{C}^{r+1} \setminus \{0\})/\sim$, where $(z_0, \ldots, z_r) \sim (z_0', \ldots, z_r')$ if and only if there exists $\lambda \in \mathbb{C}$, such that $z_i = \lambda^{n_i} z_i'$ for all $i = 0, \ldots, r$. The equivalence class of a point $(z_0, \ldots, z_r) \in \mathbb{C}^{r+1}$ under $\sim$ will be denoted by $(z_0 : \ldots : z_r)$.*

As an example, choosing $n_i = 1$ for all i, leads to the usual complex projective space $P_r(\mathbb{C})$. Weighted projective spaces are normal, irreducible, projective algebraic varieties, but in general they are singular (cf. [Do82]).

Now we have the following embeddings:

$$\mathcal{Q}_N^\circ/\!\sim \; \subset \; \mathcal{Q}_N/\!\sim \; \hookrightarrow \; P^{1,2,3,4}(\mathbb{C}),$$
$$Q(x) = x^4 + q_1 x^3 + q_2 x^2 + q_3 x + q_4 \mapsto (q_1 : q_2 : q_3 : q_4),$$
$$\mathcal{P}_N^\circ/\!\sim \; \subset \; \mathcal{P}_N/\!\sim \; \hookrightarrow \; P^{1,\dots,N-2,N}(\mathbb{C}) \quad \text{for } N > 2,$$
$$P(x) = x^N + a_1 x^{N-1} + \cdots + a_N \mapsto (a_1 : \dots : a_{N-2} : a_N),$$
$$\mathcal{P}_2^\circ/\!\sim \; \subset \; \mathcal{P}_2/\!\sim \; \hookrightarrow \; P^{2,4}(\mathbb{C}),$$
$$(P, C) \mapsto (a_2 : C),$$

where $P(x) = x^2 + a_2$ in the last map. For the Zolotarev polynomials we don't
have to care about the coefficient a_{N-1}, since we required them to be admissible, so
$P'(0) = 0$, i.e. $a_{N-1} = 0$.

The images of these embeddings will be denoted by $\mathcal{N}_N$ (resp. $\mathcal{N}_N^\circ$) for $\mathcal{Q}_N/\!\sim$
(resp. $\mathcal{Q}_N^\circ/\!\sim$) and $\mathcal{M}_N$ (resp. $\mathcal{M}_N^\circ$) for $\mathcal{P}_N/\!\sim$ (resp. $\mathcal{P}_N^\circ/\!\sim$).

The sets $\mathcal{N}_N$ and $\mathcal{M}_N$ are algebraic curves in the respective weighted projective spaces.
Their defining equations can be explicitly determined by a closer look at equation (1)
(cf. [Ju89]):

For $i \in \mathbb{Z}$ we inductively define polynomials $A_i \in \mathbb{Q}[z_1, \dots, z_4, N]$ by $A_i = 0$ for
$i < 0$, $A_0 = 1$, and

$$A_i = \frac{1}{2i(2N - i)} \sum_{k=1}^{4} (N + k - i)(2N + k - 2i) z_k A_{i-k} \quad \text{for } i > 0.$$

As one easily sees, the A_i are homogeneous of weight i (for $i > 0$) if one assigns the
weights k to z_k and 0 to N. Furthermore if $Q(x) = x^4 + q_1 x^3 + \cdots + q_4 \in \mathcal{Q}_N$ then
for $P = \Omega_N(Q) = x^N + a_1 x^{N-1} + \cdots + a_N \in \mathcal{P}_N$ we have $a_i = A_i(q_1, \dots, q_4, N)$.
From this one easily derives (cf. [Ju89]):

Proposition: *The set $\mathcal{N}_N \subset P^{1,2,3,4}(\mathbb{C})$ is equal to the algebraic curve given by*
$A_{N-1}(z_1, \dots, z_4, N) = 0$ and $A_{N+1}(z_1, \dots, z_4, N) = 0$. Thus for a polynomial
$Q(x) = x^4 + q_1 x^3 + \cdots + q_4$ there exists a Zolotarev polynomial $P(x)$ of degree N
satisfying (1), *if and only if $A_{N-1}(q_1, \dots, q_4, N) = 0$ and $A_{N+1}(q_1, \dots, q_4, N) = 0$.*

Examples: For the first few polynomials A_i we get:

$$A_1 = \frac{N}{2} q_1,$$
$$A_2 = \frac{N}{16}\left((2N - 3)q_1^2 + 4q_2\right),$$
$$A_3 = \frac{N}{96}\left((2N - 5)(N - 2)q_1^3 + 12(N - 2)q_1 q_2 + 16 q_3\right),$$
$$A_4 = \frac{N}{1536}\left((2N - 7)(2N - 5)(N - 3)q_1^4 + 24(2N - 5)(N - 3)q_1^2 q_2\right.$$
$$\left. + 32(4N - 9)q_1 q_3 + 48(N - 3)q_2^2 + 192 q_4\right).$$

The algebraic curves $\mathcal{N}_2$ to $\mathcal{N}_5$ are given by:

$$\mathcal{N}_2 = \{(q_1 : \ldots : q_4) \in P^{1,2,3,4}(\mathbb{C}) \mid q_1 = 0,\, q_3 = 0\} \cong P^{2,4}(\mathbb{C}) \cong P^{1,2}(\mathbb{C}) \cong P_1(\mathbb{C}),$$

$$\mathcal{N}_3 = \{(q_1 : \ldots : q_4) \in P^{1,2,3,4}(\mathbb{C}) \mid q_2 = -\frac{3}{4} q_1^2,\, q_4 = -\frac{1}{2} q_1 q_3\} \cong P^{1,3}(\mathbb{C}) \cong P_1(\mathbb{C}),$$

$$\mathcal{N}_4 = \{(q_1 : \ldots : q_4) \in P^{1,2,3,4}(\mathbb{C}) \mid q_3 = -\frac{3}{8} q_1 \left(q_1^2 + 4q_2\right),$$
$$5q_1^2 q_3 + 24 q_1 q_4 + 4 q_2 q_3 = 0\}$$
$$\cong \{(q_1 : q_2 : q_4) \in P^{1,2,4}(\mathbb{C}) \mid q_1 \left(\left(q_1^2 + 4q_2\right)\left(5q_1^2 + 4q_2\right) - 64 q_4\right) = 0\}$$
$$\cong P^{2,4}(\mathbb{C}) \cup P^{1,2}(\mathbb{C}) \cong P_1(\mathbb{C}) \cup P_1(\mathbb{C}),$$

where the two components intersect in the point $(0 : 2 : 0 : 1)$, and

$$\mathcal{N}_5 = \{(q_1 : \ldots : q_4) \in P^{1,2,3,4}(\mathbb{C}) \mid q_4 = -\frac{1}{96} \left(176 q_3 q_1 + 48 q_2^2 + 120 q_2 q_1^2 + 15 q_1^4\right),$$
$$6 q_4 \left(12 q_2 + 21 q_1^2\right) + 16 q_3^2 + 36 q_3 q_2 q_1 + 15 q_3 q_1^3 = 0\}$$
$$\cong \{(q_1 : q_2 : q_3) \in P^{1,2,3}(\mathbb{C}) \mid 315 q_1^6 + 2700 q_1^4 q_2 + 3456 q_1^3 q_3 + 2448 q_1^2 q_2^2$$
$$+ 1536 q_1 q_2 q_3 + 576 q_2^3 - 256 q_3^2 = 0\}.$$

Analogously one can find the equations for the coefficients of the curve $\mathcal{M}_N$ of Zolotarev polynomials (cf. [Ju89]).

Examples: The set of Zolotarev polynomials in degrees two to five is given by

$$\mathcal{M}_2 = \{(a_2 : C) \in P^{2,4}(\mathbb{C})\} = P^{2,4}(\mathbb{C}),$$
$$\mathcal{M}_3 = \{(a_1 : a_3) \in P^{1,3}(\mathbb{C})\} = P^{1,3}(\mathbb{C}),$$
$$\mathcal{M}_4 = \{(a_1 : a_2 : a_4) \in P^{1,2,4}(\mathbb{C}) \mid a_1 \left(128 a_2^2 - 144 a_2 a_1^2 + 27 a_1^4 - 512 a_4\right) = 0\},$$
$$\mathcal{M}_5 = \{(a_1 : a_2 : a_3 : a_5) \in P^{1,2,3,5}(\mathbb{C}) \mid$$
$$a_5 = \frac{1}{3125} \left(875 a_3 a_2 - 950 a_3 a_1^2 - 725 a_2^2 a_1 + 700 a_2 a_1^3 - 128 a_1^5\right),$$
$$25 a_3^2 - 70 a_3 a_2 a_1 + 4 a_3 a_1^3 - 20 a_2^3 + 37 a_1^2 a_2^2 - 8 a_2 a_1^4 = 0\}.$$

For $N > 5$ there are $N - 5$ equations that express a_5 to a_{N-2} and a_N as polynomials in a_1 to a_4. The remaining two equations involve only the variables a_1 up to a_4 and are homogeneous of degree $N - 1$ and $N + 1$.

The curves $\mathcal{N}_N$ and $\mathcal{M}_N$ are in general (i.e. for N not prime) not irreducible. In fact we already saw, that $\mathcal{N}_d$ is contained in $\mathcal{N}_N$ for each divisor d of N, and that $\mathcal{M}_d$ is embedded in $\mathcal{M}_N$. We will see later, that the curves have exactly one component for each divisor d of N and that the components only intersect in degenerate Zolotarev polynomials.

3   The differential equation — revisited

In this section we want to recall the basic constructions of Chapter 7 and Appendix I, §7, to derive a close connection between pairs (L, z_0), where $L \subset \mathbb{C}$ is a lattice and $z_0 \in \mathbb{C}/L$ is an N-division point, and Zolotarev polynomials.

We will often use such pairs (L, z_0) as above, so we define

$$\mathcal{L} := \{(L, z_0) \mid L \subset \mathbb{C} \text{ a lattice}, z_0 \in \mathbb{C}/L, z_0 \neq 0\},$$
$$\mathcal{L}_N := \{(L, z_0) \in \mathcal{L}, N \cdot z_0 = 0\}.$$

For $(L, z_0) \in \mathcal{L}_N$ there exists a well-defined function $h = h_{L,z_0,N}$ with the following properties (cf. Chapter 7 and Appendix I, §6):

1) h is elliptic for the lattice L,
2)   the divisor of h is $(h) = N \cdot (0) - N \cdot (z_0)$,
3) h is normalized by $h(z) = z^N + O(z^{N+1})$.

The function h depends on N, not only on (L, z_0), since we did not demand N to be the exact order of z_0. But one easily shows that $h_{L,z_0,nN} = (h_{L,z_0,N})^n$. In the following we will also make use of the function $g = g_{L,z_0} := \frac{1}{N} h'_{L,z_0,N}/h_{L,z_0,N}$ (which does not depend on N by the above). This function has the properties

1) g is elliptic for the lattice L,
2)   its poles are in 0 and in z_0, both of order one with residues 1 and -1 resp.

In Chapter 7 we also introduced the involution

$$\tau : z \mapsto z_0 - z$$

and we saw that g is invariant under τ: $g \circ \tau = g$. Furthermore any function u on $\mathbb{C}/L$ and invariant under τ is a rational function in g and in fact a polynomial in g if u has only poles in 0 and z_0. In Chapter 7, section 7.2, this was used to get the two differential equations

$$P(g) = Ch + \frac{1}{h},$$

and

$$Q(g) = g'^2,$$

where $P(x) = P_{L,z_0,N} = x^N + \cdots$ and $Q(x) = Q_{L,z_0}(x) = x^4 + \cdots$ are normalized polynomials of degree N resp. 4 and C is a constant depending on L, z_0, and N. The two polynomials were related by the equation

$$Q \cdot P'^2 = N^2 x^2 (P^2 - 4C),$$

so P is a normalized Zolotarev polynomial. The discriminant of the polynomial Q does not vanish (i.e. Q has only simple zeroes): From the differential equation for g we see that $Q(x)$ vanishes for $x = g(z)$ where $g'(z) = 0$. Since g has two poles of order one it takes each value exactly twice (counted with multiplicities) and its derivative takes each value four times. Now g is invariant under τ, so its derivative g' vanishes exactly at the four distinct points $z_1, \ldots, z_4$ characterized by $\tau(z_i) = z_i$. But the four values $g(z_i)$ have to be distinct, since g takes them at least with order two ($g'(z_i) = 0$). Thus Q has only simple zeroes.

Therefore the Zolotarev polynomial P is non-degenerate, and the differential equations for the elliptic functions h and g finally give us maps

$$\Psi_N : \mathcal{L}_N \to \mathcal{P}_N^\circ$$
$$(L, z_0) \mapsto \begin{cases} P_{L,z_0,N}, & \text{for } N \neq 2, \\ (P_{L,z_0,2}, C), & \text{for } N = 2, \end{cases}$$

and

$$\Phi_N : \mathcal{L}_N \to \mathcal{Q}_N^\circ$$
$$(L, z_0) \mapsto Q_{L,z_0}.$$

Since $P_{L,z_0,N}$ is a Zolotarev polynomial satisfying (1) with Q_{L,z_0}, we obviously have $\Psi_N = \Omega_N \circ \Phi_N$.

If z_0 is an N-division point of $\mathbb{C}/L$, then we can also take it as an nN-division point, i.e. $\mathcal{L}_N \subset \mathcal{L}_{nN}$. The function g does not depend on N as well as the polynomial Q, so we have

$$P_{L,z_0,nN}(x) = \iota_n(P_{L,z_0,N}(x))$$

and

$$C_{L,z_0,nN} = C_{L,z_0,N}^n,$$

where ι_n is the injection from the first section.

Now we can extend the commutative diagram of section 1 to:

$$\Psi_N \begin{pmatrix} \mathcal{L}_N & \subset & \mathcal{L}_{nN} \\ \Phi_N \downarrow & & \downarrow \Phi_{nN} \\ \mathcal{Q}_N^\circ & \subset & \mathcal{Q}_{nN}^\circ \\ \Omega_N \downarrow & & \downarrow \Omega_{nN} \\ \mathcal{P}_N^\circ & \overset{\iota_n}{\hookrightarrow} & \mathcal{P}_{nN}^\circ \end{pmatrix} \Psi_{nN}$$

4 Modular interpretation of Zolotarev polynomials

In the last section we saw that we can associate a non-degenerate Zolotarev polynomial $P_{L,z_0,N} = \Psi_N(L, z_0)$ to each pair $(L, z_0) \in \mathcal{L}_N$. Now we want to prove the

Theorem: *The maps* Φ_N *and* Ψ_N *are bijections.*

Proof: Since $\Psi_N = \Omega_N \circ \Phi_N$ and Ω_N is a bijection, it suffices to proof bijectivity for the map Φ_N. For this we need the following

Proposition: *Let* $Q(x) = x^4 + q_1 x^3 + q_2 x^2 + q_3 x + q_4$ *be a normalized polynomial of degree 4, such that* $\operatorname{discr}(Q) \neq 0$. *Then there exists a unique meromorphic function* $g(z)$ *with* $g(z) = \frac{1}{z} + O(1)$ *and* $Q(g) = g'^2$. *This function is elliptic for a lattice* $L \subset \mathbb{C}$ *and with* $\widetilde{Q}(x) = Q(x - q_1/4) = x^4 + \tilde{q}_2 x^2 + \tilde{q}_3 x + \tilde{q}_4$ *we have*

$$g(z) = -\frac{1}{2} \cdot \frac{\wp'(z) + \wp'(z_0)}{\wp(z) - \wp(z_0)} - \frac{q_1}{4}$$

where $\wp$ *is the Weierstraß* $\wp$*-function for the lattice* L *with lattice constants* $g_2(L) = \tilde{q}_4 + \tilde{q}_2^2/12$ *and* $g_3(L) = \tilde{q}_4 \tilde{q}_2/6 - \tilde{q}_3^2/16 - \tilde{q}_2^3/216$ *and* z_0 *is the point on* $\mathbb{C}/L$ *with* $\wp(z_0) = -\tilde{q}_2/6$ *and* $\wp'(z_0) = \tilde{q}_3/4$. 📕

For the proof of this proposition we refer the reader to [Ju89] (see also [WhWa69], pp. 452–455).

The injectivity of Φ_N now immediately follows from the proposition: If $\Phi_N(L, z_0) = Q_{L,z_0}(x) = x^4 + q_1 x^3 + \cdots$, then we have $Q_{L,z_0}(g) = g'^2$ for a uniquely determined function g, so $g = g_{L,z_0}$. But then the lattice L is the periodicity lattice of g and z_0 is the pole of g on $\mathbb{C}/L$ that is different from zero. Therefore (L, z_0) is uniquely determined by $Q_{L,z_0} = \Phi_N(L, z_0)$, i.e. Φ_N is injective.

To prove the surjectivity of Φ_N we again use the proposition: Let $Q \in \mathcal{Q}_N^\circ$, so especially $\operatorname{discr}(Q) \neq 0$. Let $g(z)$ be the function in the proposition, i.e. $g'^2 = Q(g)$ and g is elliptic for some lattice L. From the explicit formula for g in the proposition we see that g has only two poles, namely one in 0 and one in z_0, both of order one with residues 1 resp. -1. Now we have to show that z_0 is an N-division point of $\mathbb{C}/L$ and that $g = g_{L,z_0}$.

Since $Q \in \mathcal{Q}_N^\circ$ we know that $P := \Omega_N(Q)$ (resp. $(P, C) = \Omega_2(Q)$ for $N = 2$) is a normalized Zolotarev polynomial with $QP'^2 = N^2 x^2 (P^2 - A^2)$ (and $A^2 = 4C$ for $N = 2$). Denote by f_1 and f_2 the functions $f_1 := P(g)$ and $f_2 := \frac{1}{N} f_1'/g = \frac{1}{N} P'(g) g'/g$ and by h the function $h := \frac{2}{A^2}(f_1 + f_2)$. All of them are elliptic for L

since g is. From (1) and the definitions we have

$$f_2^2 = \left(\frac{1}{Ng}P'(g)g'\right)^2 = \left(\frac{1}{Ng}P'(g)\right)^2 Q(g) =$$
$$= P(g)^2 - A^2 = f_1^2 - A^2$$
$$\Rightarrow \quad 2f_2 f_2' = 2f_1 f_1'$$
$$\Rightarrow \quad f_2' = \frac{f_1'}{f_2}f_1 = Ngf_1.$$

For the logarithmic derivative of h we get

$$\frac{h'}{h} = \frac{f_1' + f_2'}{f_1 + f_2} = \frac{f_1' + Ngf_1}{f_1 + \frac{1}{Ng}f_1'} = Ng. \tag{2}$$

Since the only poles of g are in 0 and z_0 of order one and with residues 1 resp. -1 we get from (2) for the divisor (h) of h:

$$(h) = N \cdot (0) - N \cdot (z_0)$$
$$\Rightarrow \quad N \cdot 0 - N \cdot z_0 = 0 \in \mathbb{C}/L$$
$$\Rightarrow \quad N \cdot z_0 = 0 \in \mathbb{C}/L.$$

Therefore z_0 is an N-division point of $\mathbb{C}/L$. Since h has the same divisor as $h_{L,z_0,N}$ and

$$\frac{1}{h} = \frac{A^2}{2(f_1 + f_2)} = \frac{A^2(f_1 - f_2)}{2(f_1^2 - f_2^2)} =$$
$$= \frac{1}{2}(f_1 - f_2) = \frac{1}{z^N} + O\left(\frac{1}{z^{N-1}}\right)$$

satisfies the normalization condition, we have $h = h_{L,z_0,N}$ and (2) shows that $g = g_{L,z_0}$. Since $Q(g) = g'^2$ we finally get $Q = Q_{L,z_0} = \Phi_N(L,z_0)$, so Φ_N is surjective. ☐

In the next section degenerate Zolotarev polynomials will show up as values of Ψ_N on degenerate lattices.

5   The embedding of the modular curve

In Appendix I, §7, we already saw that two pairs (L, z_0) and (L', z_0') in $\mathcal{L}_N$ are isomorphic ($(L, z_0) \sim (L', z_0')$), if and only if $(L', z_0') = \lambda(L, z_0)$ for some $\lambda \in \mathbb{C}^*$, where the action of $\mathbb{C}^*$ on $\mathcal{L}_N$ is given by $\lambda(L, z_0) = (\lambda L, \lambda z_0)$. As one can easily verify using the axiomatic definition of the function h_{L,z_0}, we have

$$h_{\lambda L, \lambda z_0, N}(z) = \lambda^N h_{L,z_0,N}(z/\lambda)$$

and so

$$g_{\lambda L, \lambda z_0}(z) = g_{L, z_0}(z/\lambda)/\lambda.$$

Therefore $Q_{\lambda L, \lambda z_0} = \lambda(Q_{L, z_0})$ and $P_{\lambda L, \lambda z_0, N} = \lambda(P_{L, z_0, N})$, and the maps Φ_N and Ψ_N are equivariant with respect to the $\mathbb{C}^*$-actions on $\mathcal{L}_N$ and $\mathcal{Q}_N$ resp. $\mathcal{P}_N$ (cf. section 2). Since they were bijections, they induce bijections on the set of equivalence classes:

$$
\begin{array}{ccccc}
\Phi_N & \mathcal{Q}_N^\circ/\sim & \overset{\cong}{\to} & \mathcal{N}_N^\circ & \subset \quad P^{1,2,3,4}(\mathbb{C}) \\
& \nearrow & & & \\
\mathcal{L}_N/\sim & & \cong \downarrow \Omega_N & & \\
& \searrow & & & \\
\Psi_N & \mathcal{P}_N^\circ/\sim & \overset{\cong}{\to} & \mathcal{M}_N^\circ & \subset \quad P^{1,\dots,N-2,N}(\mathbb{C})
\end{array}
$$

If we define $\mathcal{L}_{N,\mathrm{p}} = \{(L, z_0) \in \mathcal{L}_N \mid z_0$ has exact order N in $\mathbb{C}/L\}$ then the set $\mathcal{L}_N$ can be naturally written as the disjoint union $\mathcal{L}_N = \bigcup_{d \mid N} \mathcal{L}_{d,\mathrm{p}}$. This splitting is respected by the $\mathbb{C}^*$-action on $\mathcal{L}_N$. Furthermore for any divisor d of N we define $\mathcal{Q}_{d,\mathrm{p}}^\circ := \Phi_N(\mathcal{L}_{d,\mathrm{p}}) = \Phi_d(\mathcal{L}_{d,\mathrm{p}})$ and $\mathcal{P}_{d,N,\mathrm{p}}^\circ := \Psi_N(\mathcal{L}_{d,\mathrm{p}}) = \iota_{N/d}(\mathcal{P}_{d,d,\mathrm{p}}^\circ)$. Then we have $\mathcal{Q}_N^\circ = \bigcup_{d \mid N} \mathcal{Q}_{d,\mathrm{p}}^\circ$ and $\mathcal{P}_N^\circ = \bigcup_{d \mid N} \mathcal{P}_{d,N,\mathrm{p}}^\circ$. Finally let $\mathcal{N}_{d,\mathrm{p}}^\circ \subset \mathcal{N}_N^\circ$ (resp. $\mathcal{M}_{d,N,\mathrm{p}}^\circ \subset \mathcal{M}_N^\circ$) be the images of $\mathcal{Q}_{d,\mathrm{p}}^\circ$ (resp. $\mathcal{P}_{d,N,\mathrm{p}}^\circ$) under the embeddings into weighted projective spaces we discussed in section 2.

From Appendix I, §7.1, we know that the set of equivalence classes $\mathcal{L}_{d,\mathrm{p}}/\sim$ can be parametrized by $\Gamma_1(d)\backslash\mathfrak{h}$ via $\tau \mapsto (2\pi i(\mathbb{Z}\tau + \mathbb{Z}), 2\pi i/d)$. Restricting the above diagram to primitive d-division points, we have

$$
\begin{array}{ccccc}
& & \Phi_N & \mathcal{N}_{d,\mathrm{p}}^\circ & \subset \quad P^{1,2,3,4}(\mathbb{C}) \\
& & \nearrow & & \\
\Gamma_1(d)\backslash\mathfrak{h} & \overset{\cong}{\to} & \mathcal{L}_{d,\mathrm{p}}/\sim & & \\
& & \searrow & & \\
& & \Psi_N & \mathcal{M}_{d,N,\mathrm{p}}^\circ & \subset \quad P^{1,\dots,N-2,N}(\mathbb{C})
\end{array}
\qquad (3)
$$

Thus we get two embeddings of $\Gamma_1(d)\backslash\mathfrak{h}$ into weighted projective spaces with images given by $\mathcal{N}_{d,\mathrm{p}}^\circ$ and $\mathcal{M}_{d,N,\mathrm{p}}^\circ$. From Appendix I, §7.1, we know, that in the homogeneous coordinates of the weighted projective spaces the embeddings are given by modular forms. To extend the maps to the whole of $\overline{\Gamma_1(d)\backslash\mathfrak{h}}$ (i.e. the union of $\Gamma_1(d)\backslash\mathfrak{h}$ and the cusps), we have to know their values on the cusps of $\Gamma_1(d)$. From Chapter 7, section 7.6 we know that the values of Φ_N on the cusps of $\Gamma_1(d)$ are given by

(i) $\quad Q(x) = (x - k/d)(x + (d - k)/d)$ for $0 < k < d,$

(ii) $\quad Q(x) = x^2\big(x^2 + 2\,\frac{1+\zeta^l}{1-\zeta^l}\,x + 1\big)$ for $\zeta = e^{\frac{2\pi i}{d}}$ and $\gcd(l, d) = 1.$

Thus the values of Φ_N on the cusps are polynomials Q that correspond to degenerate Zolotarev polynomials of degree N and we can combine the maps in (3), extended to the compactified modular curve $\overline{\Gamma_1(d)\backslash\mathfrak{h}}$, to get

$$
\begin{array}{ccc}
 & \overset{\overline{\Phi_N}}{} \mathcal{N}_N & \subset \quad P^{1,2,3,4}(\mathbb{C}) \\
 & \nearrow & \\
\overset{\cdot}{\bigcup_{d\mid N}} \overline{\Gamma_1(d)\backslash\mathfrak{h}} & & \\
 & \searrow & \\
 & \overset{\overline{\Psi_N}}{} \mathcal{M}_N & \subset \quad P^{1,\ldots,N-2,N}(\mathbb{C})
\end{array}
$$

The maps $\overline{\Phi_N}$ and $\overline{\Psi_N}$ are surjective and away from the cusps they are also injective. The degenerate Zolotarev polynomials of type (ii) have only one preimage, the ones of type (i) have $\varphi(\gcd(k,d))$ preimages on the component $\overline{\Gamma_1(d)\backslash\mathfrak{h}}$, where φ is the Euler φ-function. Therefore the maps are embeddings away from the cusps and are also embeddings for components $\overline{\Gamma_1(d)\backslash\mathfrak{h}}$ with d a prime number. Furthermore, the images of the different components only intersect in degenerate Zolotarev polynomials of type (i).

6   Applications to elliptic genera

The image of a genus on the rational cobordism ring is equal to the $\mathbb{Q}$-algebra Λ generated by the coefficients of the characteristic power series $x/f(x)$, which in turn is equal to the $\mathbb{Q}$-algebra generated by the coefficients of the power series f'/f. In the case of elliptic genera of level N we have $f'/f = g$, where g satisfies a differential equation $g'^2 = Q(g)$ with $Q \in \mathcal{Q}_N^\circ$. So for $Q(x) = x^4 + q_1 x^3 + q_2 x^2 + q_3 x + q_4$ we have $\Lambda = [q_1,\ldots,q_4]$. Since $Q(x) \in \mathcal{Q}_N^\circ$ the coefficients of the polynomial satisfy the two relations $A_{N-1}(q_1,\ldots,q_4,N) = 0$ and $A_{N+1}(q_1,\ldots,q_4,N) = 0$, we described in section 2 of this appendix.

For the universal elliptic genus of level N we also know that q_1 up to q_4 are modular forms on $\Gamma_1(N)$, so we get

Proposition: *The rational image Λ of the universal elliptic genus of level N is generated by 4 modular forms $q_1,\ldots,q_4$, of weight 1 up to 4, which satisfy 2 weighted homogeneous relations $A_{N-1}(q_1,\ldots,q_4,N) = 0$ and $A_{N+1}(q_1,\ldots,q_4,N) = 0$, of degree $N-1$ and $N+1$.*

For general N there will be more relations, since the algebraic curve with function field $\mathbb{Q}[q_1,\ldots,q_4]/(A_{N-1},A_{N+1})$ has one component for each divisor d of N, the component parametrized by modular forms on $\Gamma_1(N)$ is the one with $d = N$. For N prime, we have only one component and there are no more relations.

That means for large N the image of the genus will only be a small subring of the ring of modular forms on $\Gamma_1(N)$.

In his Diplomarbeit [Hö91], G. Höhn studies a universal elliptic genus φ_{ell} that generalizes elliptic genera of all levels:

This genus belongs to the characteristic power series $x/f(x)$ such that $g = f'/f$ satisfies $g'^2 = Q(g)$ for $Q(x) = x^4 + q_1 x^3 + q_2 x^2 + q_3 x + q_4$ where $q_1,\ldots,q_4$ are indeterminates, i.e. the polynomial Q is no longer restricted to belong to a Zolotarev polynomial. Elliptic genera of any level N factorize over this genus φ_{ell}. After a homogeneous change of coordinates Höhn gets

$$\varphi_{\mathrm{ell}} : \Omega \otimes \mathbb{Q} \to \mathbb{Q}[A, B, C, D],$$

where A,B,C,D are again indeterminates of weights 1 up to 4 (i.e. q_1 up to q_4 can be expressed as homogeneous polynomials in A, B, C, D and vice versa), and for an SU manifold M, i.e. a stably almost complex manifold M with $c_1(M) = 0$, there holds $\varphi_{\mathrm{ell}}(M) \in \mathbb{Q}[B, C, D]$.

Using the rigidity theorem of Hirzebruch for level N genera (see Appendix III), Höhn shows the following

Theorem: *The universal elliptic genus φ_{ell} is rigid for S^1-actions on SU manifolds.* ▱

Another result he proves is the invariance of level N elliptic genera under blow ups in codimension 1 modulo N. Furthermore Höhn determines the kernels of φ_{ell} and of the level N elliptic genera in terms of ideals of manifolds with certain kinds of S^1-actions and ideals generated by certain fibre bundles.

Bibliography

[Ap76]   T.M. Apostol: *Modular Functions and Dirichlet Series in Number Theory.* Graduate Texts in Mathematics; 41. Berlin-Heidelberg-New York, Springer, 1976.

[AtBo67]   M.F. Atiyah, R. Bott: A Lefschetz fixed point formula for elliptic complexes: I, *Ann. of Math.* **86** (1967), pp. 374–407.

[AtBoSh64]   M.F. Atiyah, R. Bott, A. Shapiro: Clifford modules, *Topology* **3**, suppl. 1 (1964), pp. 3–38.

[AtHi59a]   M.F. Atiyah, F. Hirzebruch: Quelques théorèmes de non-plongement pour les variétés différentiables, *Bull. Soc. Math. France* **87** (1959), pp. 383–396.

[AtHi59b]   M.F. Atiyah, F. Hirzebruch: Riemann-Roch theorems for differentiable manifolds, *Bull. Amer. Math. Soc.* **65** (1959), pp. 276–281.

[AtHi61a]   M.F. Atiyah, F. Hirzebruch: Cohomologie-Operationen und charakteristische Klassen, *Math. Z.* **77** (1961), pp. 149–187.

[AtHi61b]   M.F. Atiyah, F. Hirzebruch: Bott Periodicity and the Parallelizability of the spheres, *Proc. Cambridge Philos. Soc.* **57** (1961), pp. 223–226.

[AtHi70]   M.F. Atiyah, F. Hirzebruch: Spin-Manifolds and Group Actions; pp. 18–28 in: *Essays on Topology and Related Topics. Memoires dédiés à Georges de Rham.* Berlin-Heidelberg-New York, Springer, 1970.

[AtSe68II]   M.F. Atiyah, G.B. Segal: The index of elliptic operators: II, *Ann. of Math.* **87** (1968), pp. 531–545.

[AtSi68I]   M.F. Atiyah, I.M. Singer: The index of elliptic operators: I, *Ann. of Math.* **87** (1968), pp. 484–530.

[AtSi68III]   M.F. Atiyah, I.M. Singer: The index of elliptic operators: III, *Ann. of Math.* **87** (1968), pp. 546–604.

[BoHi58]   A. Borel, F. Hirzebruch: Characteristic classes and homogeneous spaces, I, *Amer. J. Math.* **80** (1958), pp. 458–538.

[BoHi59]   A. Borel, F. Hirzebruch: Characteristic classes and homogeneous spaces, II, *Amer. J. Math.* **81** (1959), pp. 315–382.

[BoTa89]   R. Bott, C.H. Taubes: On the rigidity theorems of Witten, *J. Amer. Math. Soc.* **2** (1989), pp. 137–186.

[BoTu82]   R. Bott, L.W. Tu: *Differential Forms in Algebraic Topology.* Graduate Texts in Mathematics; 82. Berlin-Heidelberg-New York, Springer, 1982.

[Br72] W. Browder: *Surgery on Simply-Connected Manifolds.* Ergebnisse der Mathematik und ihrer Grenzgebiete; Band 65. Berlin-Heidelberg-New York, Springer, 1972.

[Bu70] В.М. Бухштабер: Характер Чженя-Дольда в кобордизмах. I, *Mat. Sb.* **83** (**125**) (1970), pp. 575–595. English translation: V.M. Bukhshtaber: The Chern-Dold Character in Cobordisms. I, *Math. USSR-Sb.* **12** (1970), pp. 573–594.

[ChHiSe57] S.S. Chern, F. Hirzebruch: J-P. Serre, On the index of a fibered manifold, *Proc. Amer. Math. Soc.* **8** (1957), pp. 587–596.

[ChCh88] D.V. Chudnovsky, G.V. Chudnovsky: Elliptic Modular Functions and Elliptic Genera, *Topology* **27** (1988), pp. 163–170.

[Do82] I. Dolgachev: Weighted projective varieties; pp. 34–71 in: J.B. Carrell (Ed.): *Group Actions and Vector Fields, Proceedings 1981.* Lecture Notes in Mathematics; 956. Berlin-Heidelberg-New York, Springer, 1982.

[EiZa85] M. Eichler, D. Zagier: *The Theory of Jacobi Forms.* Progress in Mathematics; Vol. 55. Boston-Basel-Stuttgart, Birkhäuser, 1985.

[Eu251] L. Euler: De integratione aequationis differentialis $\frac{m\,dx}{\sqrt{(1-x^4)}} - \frac{n\,dy}{\sqrt{(1-y^4)}}$, Commentatio 251 indicis Enestroemiani, *Novi commentarii academiae scientiarum Petropolitanae* **6** (1756/7), 1761, pp. 37–57, Summarium ibidem, pp. 7–9. Or: Opera Omnia, Series I: Opera mathematica, Volumen XX: Commentationes analyticae ad theoriam integralium ellipticorum pertinentes, Volumen prius, Ed. Adolf Krazer, Lipsiae et Berolini, B.G. Teubner, MCMXII, reprint 1978, pp. 58–79.

[Gu62] R.C. Gunning: *Lectures on Modular Forms*, Notes by A. Brumer. Annals of Mathematics Studies; 48. Princeton, Princeton University Press, 1962.

[Ha78] A. Hattori: Spin^c-Structures and S^1-Actions, *Invent. Math.* **48** (1978), pp. 7–31.

[Hi53] F. Hirzebruch: Über die quaternionalen projektiven Räume, *S.-B. Math.-Nat. Kl. Bayer. Akad. Wiss.* (1953), pp. 301–312.

[Hi54] F. Hirzebruch: Der Satz von Riemann-Roch in faisceau-theoretischer Formulierung. Einige Anwendungen und offene Fragen; pp. 457–473 in: *Proc. Int. Congr. Math. 1954*, Vol. III. Groningen, Noordhoff, and Amsterdam, North-Holland Publishing Co., 1956.

[Hi56] F. Hirzebruch: *Neue topologische Methoden in der algebraischen Geometrie.* Ergebnisse der Mathematik und ihrer Grenzgebiete; Heft 9. Berlin-Heidelberg, Springer, 1956. English translation: *Topological Methods in Algebraic Geometry* (3rd edition, with appendices by R.L.E. Schwarzenberger and A. Borel). Grundlehren der mathematischen Wissenschaften; 131. Berlin-Heidelberg, Springer, 1966.

[Ho50]   W.V.D. Hodge: The Topological Invariants of Algebraic Varieties; pp. 182–192 in: *Proc. Int. Congr. Math. 1950*, vol. I. Providence, AMS, 1952.

[Hö91]   G. Höhn: *Komplexe elliptische Geschlechter und S^1-äquivariante Kobordismustheorie*. Diplomarbeit, Bonn, 1991.

[HoSa40]   H. Hopf, H. Samelson: Ein Satz über die Wirkungsräume geschlossener Liescher Gruppen, *Commen. Math. Helv.* **13** (1940/41), pp. 240–251.

[Ju89]   R. Jung: *Zolotarev-Polynome und die Modulkurve $X_1(N)$*. Diplomarbeit, Bonn, 1989.

[KeMi63]   M.A. Kervaire, J.W. Milnor: Groups of homotopy spheres: I, *Ann. of Math.* **77** (1963), pp. 504–537.

[Ko67]   K. Kodaira: A certain type of irregular algebraic surfaces, *J. Analyse Math.* **19** (1967), pp. 207–215.

[Ko70]   C. Kosniowski: Applications of the holomorphic Lefschetz formula, *Bull. London Math. Soc.* **2** (1970), pp. 43–48.

[La86]   P.S. Landweber: Elliptic Cohomology and Modular Forms; pp. 55–68 in [La88].

[La88]   P.S. Landweber (Ed.): *Elliptic Curves and Modular Forms in Algebraic Topology, Proceedings Princeton 1986*. Lecture Notes in Mathematics; 1326. Berlin-Heidelberg, Springer, 1988.

[LaSt88]   P.S. Landweber, R.E. Stong: Circle Actions on Spin Manifolds and Characteristic Numbers, *Topology* **27** (1988), pp. 145–161.

[La73]   S. Lang: *Elliptic functions*. Reading, Addison Wesley, 1973. Now: Graduate Texts in Mathematics; 112. Berlin-Heidelberg-New York, Springer, 1987.

[Lu71]   G. Lusztig: Remarks on the holomorphic Lefschetz numbers; pp. 193–204 in: *Analyse Globale*. Sém. Math. Sup.; été 1969. Montréal, Les Presses de l'Université de Montréal, 1971.

[Ma79]   А.И. Маркушевич: *Введение в классическую теорию абелевых функций*. Moscow, Nauka, 1979. English translation: A.I. Markushevich: *Introduction to the classical theory of abelian functions*. Translations of Mathematical Monographs; Vol. 96. Providence, AMS, 1992.

[Mi58]   J. Milnor: Some consequences of a theorem of Bott, *Ann. of Math.* **68** (1958), pp. 444–449.

[Mi65]   J. Milnor: On the Stiefel-Whitney numbers of complex manifolds and of spin manifolds, *Topology* **3** (1965), pp. 223–230.

[Mi ∞]   J. Milnor: On the cobordism ring Ω^* and a complex analogue, part II; unpublished.

[MiKe58]   J. Milnor, M.A. Kervaire: Bernoulli Numbers, Homotopy Groups and a Theorem of Rohlin; pp. 454–458 in: *Proc. Int. Congr. Math. 1958*. Cambridge, Cambridge University Press, 1960.

[MiSt74]   J. Milnor, J.B. Stasheff: *Characteristic Classes*. Annals of Mathematics Studies; 76. Princeton, Princeton University Press, 1974.

[No62]   С.П. Новиков: Гомотопические свойства комплексов Тома (S.P. Novikov: Homotopy Properties of Thom Complexes), *Mat. Sb.* **57** (**99**) (1962), pp. 407–442

[Oc81]   S. Ochanine: *Signature modulo* 16, *invariants de Kervaire généralisés, et nombres caractéristiques dans la K-théorie réelle*. Mémoire de la Société Mathématique de France; Nouvelle Série N° 5 (Suppl. au *Bull. Soc. Math. France* **109** (1981)).

[Oc86]   S. Ochanine: Genres Elliptiques Équivariantes; pp. 107–122 in [La88].

[Oc87]   S. Ochanine: Sur les genres multiplicatifs définis par des intégrales elliptiques, *Topology* **26** (1987), pp. 143–151.

[Og69]   A. Ogg: *Modular Forms and Dirichlet Series*. Mathematics Lecture Note Series; 39. New York, Benjamin, 1969.

[Pa65]   R.S. Palais, *Seminar on the Atiyah-Singer Index Theorem*. Annals of Mathematics Studies; 57. Princeton, Princeton University Press, 1965.

[Qu69]   D. Quillen: On the formal group laws of unoriented and complex cobordism theory, *Bull. Amer. Math. Soc.* **75** (1969), pp. 1293–1298.

[Ra77]   R.A. Rankin: *Modular forms and functions*. Cambridge-London-New York-Melbourne, Cambridge University Press, 1977.

[Ro73]   A. Robert: *Elliptic Curves*. Lecture Notes in Mathematics; 326. Berlin-Heidelberg-New York, Springer, 1973.

[Sch74]   B. Schoeneberg: *Elliptic Modular Functions*. Grundlehren der mathematischen Wissenschaften; 203. Berlin-Heidelberg-New York, Springer, 1974.

[Se70]   J-P. Serre: *Cours d'Arithmétique*. Paris, Presses Universitaires de France, 1970. English translation: *A Course in Arithmetic*. Graduate Texts in Mathematics; 7. Berlin-Heidelberg-New York, Springer, 1973.

[Sh71]   G. Shimura: *Introduction to the Arithmetic Theory of Automorphic Functions*. Publications of the Mathematical Society of Japan; 11. Iwanami Shoten, Publishers and Princeton University Press, 1971.

[Sm60]   S. Smale: The generalized Poincaré conjecture in higher dimensions, *Bull. Amer. Math. Soc.* **66** (1960), pp. 373–375.

[StEp62]   N.E. Steenrod: D.B.A. Epstein, *Cohomology Operations*. Annals of Mathematics Studies; 50. Princeton, Princeton University Press, 1962.

[St68]   R.E. Stong: *Notes on Cobordism Theory*. Mathematical Notes; 7. Princeton, Princeton University Press, 1968.

[Ta88]   C.H. Taubes: S^1 Actions and Elliptic Genera, *Comm. Math. Phys.* **122** (1989), pp. 455–526.

[Th53] R. Thom: Sous-variétés et classes d'homologie des variétés différentiables I, II, *C. R. Acad. Sci. Paris Sér. I Math.* **236** (1953), pp. 453–454 and pp. 573–575.

R. Thom: Sur un problème de Steenrod, Ibid., **236** (1953), pp. 1128–1130.

R. Thom: Variétés différentiables cobordantes, Ibid., **236** (1953), pp. 1733–1735.

[Th54] R. Thom: Quelques propriétés globales des variétés différentiables, *Commen. Math. Helv.* **28** (1954), pp. 17–86.

[Wa60] C.T.C. Wall: Determination of the Cobordism Ring, *Ann. of Math.* **72** (1960), pp. 292–311.

[We80] R.O. Wells: *Differential Analysis on Complex Manifolds*. Graduate Texts in Mathematics; 65. Berlin-Heidelberg-New York, Springer, 1980.

[WhWa69] E.T. Whittaker, G.N. Watson: *A Course of MODERN ANALYSIS* (Reprint of the Fourth Edition 1927). Cambridge, Cambridge University Press, 1969.

[Wi86] E. Witten: The Index of the Dirac Operator in Loop Space; pp. 161–181 in [La88].

[Wu50] Wu Wen-Tsün: Classes caractéristiques et i-carrés d'une variété, *C. R. Acad. Sci. Paris Sér I Math.* **230** (1950), pp. 508–511.

[Za86] D. Zagier: Note on the Landweber-Stong Elliptic Genus; pp. 216–224 in [La88].

Index

Symbols

Re, real part of a complex number, 4

ρ, representation, 9

ρE, associated vector bundle, 9

$s(M)$, Milnor number of M, 42

$s_k(L)$, homogeneous lattice function, 127

$\mathrm{sgn}(s)$, sign of a permutation, 46

$\mathrm{sign}(M)$, signature of M, 41

$\mathrm{sign}(X,W)$, twisted signature, 74

$\mathrm{sign}(g,X)$, equivariant signature, 68

$\mathrm{sign}(q,\mathcal{L}X)$, formal equivariant signature of loop space, 74

$S^k E$, symmetric power of a bundle, 9

$S_t E$, formal sum of symmetric powers, 12

$S_k(\Gamma)$, space of cusp forms, 123

$\mathrm{SL}_2(\mathbb{R})$, unimodular real 2×2-matrices, 121

Sq^i, Steenrod square, 114

σ_r, elementary symmetric function, 7

$\sigma_r(n)$, number theoretic function, 129

$\sigma(x), \sigma(\tau,x), \sigma_L(x)$, Weierstraß σ-function, 82, 144

$\sigma^{(p)}(D)$, p-symbol of D, 57

td, Todd genus, 59

T_n, Todd polynomial, 20

T_N, Chebyshev polynomial, 105

$\widetilde{T}_N$, normalized Chebyshev polynomial, 105

TM, tangent bundle of M, 35

$\theta(\tau), \theta_L(\tau)$, theta series, 126, 126

Θ_L, group of theta functions for L, 143

$\widetilde{\Theta}_L$, group of trivial theta functions for L, 143

$(u_1, \ldots, u_r)$, virtual submanifold, 36

U_i, Wu class, 114

$\Upsilon(x)$, special theta function, 171

X^Δ, bundle along the fibres, 47

X^g, fixed point manifold, 66

X_ν^g, component of fixed point manifold, 66

$X_1(N)$, special Riemann surface, 158

$\zeta(s)$, Riemann ζ-function, 131

Edited by Klas Diederich

*A Publication of the Max-Planck-Institut für Mathematik, Bonn

MIX
Papier aus verantwortungsvollen Quellen
Paper from responsible sources
FSC® C105338

FSC
www.fsc.org

If you have any concerns about our products,
you can contact us on
ProductSafety@springernature.com

In case Publisher is established outside the EU,
the EU authorized representative is:
Springer Nature Customer Service Center GmbH
Europaplatz 3, 69115 Heidelberg, Germany

Printed by Libri Plureos GmbH
in Hamburg, Germany